MICRO 2016. Fate and Impact of Microplastics in Marine Ecosystems

MICRO 2016. Fate and Impact of Microplastics in Marine Ecosystems

From the Coastline to the Open Sea

Juan Baztan

Bethany Jorgensen

Sabine Pahl

Richard C. Thompson

Jean-Paul Vanderlinden

25TH, 26TH AND 27TH MAY 2016 MICRO2016 INTERNATIONAL CONFERENCE
LANZAROTE

ELSEVIER

AMSTERDAM • BOSTON • HEIDELBERG • LONDON • NEW YORK • OXFORD
PARIS • SAN DIEGO • SAN FRANCISCO • SINGAPORE • SYDNEY • TOKYO

Elsevier
Radarweg 29, PO Box 211, 1000 AE Amsterdam, Netherlands
The Boulevard, Langford Lane, Kidlington, Oxford OX5 1GB, United Kingdom
50 Hampshire Street, 5th Floor, Cambridge, MA 02139, United States

British Library Cataloguing-in-Publication Data
A catalogue record for this book is available from the British Library

Library of Congress Cataloging-in-Publication Data
A catalog record for this book is available from the Library of Congress

ISBN: 978-0-12-812271-6

For Information on all Elsevier publications
visit our website at https://www.elsevier.com

Working together
to grow libraries in
developing countries

www.elsevier.com • www.bookaid.org

Publisher: Cathleen Sether
Acquisition Editor: Louisa Hutchins
Editorial Project Manager: Emily Thomson
Production Project Manager: Priya Kumaraguruparan
Designer: MPS

Typeset by MPS Limited, Chennai, India

MICRO 2016 is an initiative of the Lanzarote Biosphere Reserve within the UNESCO MAB programme and the network of scientists Marine Sciences For Society.

Lanzarote, May 2016

Hosted by:

Under the patronage of UNESCO

CONTENTS

PART I ABSTRACTS FROM ORAL PRESENTATIONS

A. van Oyen, S. Kühn, J.A. van Franeker, M. Ortlieb and
M. Egelkraut-Holtus

M. Boertien and E. Butter

A.M. Mahon, B. O' Connell, M.G. Healy, I. O' praConnor, R. Officer,
R. Nash and L. Morrison

F. Faure, O. Pompini and L.F. de Alencastro

T. Maes and H. Leslie

T. Gundry, B. Clarke and A.M. Osborn

A.A. Horton, C. Svendsen, R.J. Williams, D.J. Spurgeon and E. Lahive

*For further information on this study, please contact the authors.

*For further information on this study, please contact the authors.

---

*For further information on this study, please contact the authors.

*For further information on this study, please contact the authors.

*For further information on this study, please contact the authors.

*For further information on this study, please contact the authors.

PART II ABSTRACTS FROM POSTERS

---

*For further information on this study, please contact the authors.

*For further information on this study, please contact the authors.

*For further information on this study, please contact the authors.

*For further information on this study, please contact the authors.

*For further information on this study, please contact the authors.

*For further information on this study, please contact the authors.

*For further information on this study, please contact the authors.

*For further information on this study, please contact the authors.

PART III LANZAROTE DECLARATION, JUNE 21 2016

PART IV BREAKING DOWN THE PLASTIC AGE

J. Baztan, M. Bergmann, A. Booth, E. Broglio, A. Carrasco, O. Chouinard,
M. Clüsener-Godt, M. Cordier, A. Cozar, L. Devrieses, H. Enevoldsen,
R. Ernsteins, M. Ferreira-da-Costa, M-C. Fossi, J. Gago, F. Galgani,
J. Garrabou, G. Gerdts, M. Gomez, A. Gómez-Parra, L. Gutow, A. Herrera,
C. Herring, T. Huck, A. Huvet, J-A. Ivar do Sul, B. Jorgensen, A. Krzan,
F. Lagarde, A. Liria, A. Lusher, A. Miguelez, T. Packard, S. Pahl,
I. Paul-Pont, D. Peeters, J. Robbens, A-C. Ruiz-Fernández, J. Runge,
A. Sánchez-Arcilla, P. Soudant, C. Surette, R.C. Thompson, L. Valdés,
J-P. Vanderlinden and N. Wallace

PART V WHERE NEXT? THE ROAD TO MICRO 2018 AND BEYOND

R.C. Thompson

*For further information on this study, please contact the authors.

We are pleased to present the scientific community and stakeholders with the proceedings from MICRO 2016, a three-day international conference, hosted in Lanzarote, Canary Islands, Spain May 25–27 2016:

MICRO 2016. Fate and Impact of Microplastics in Marine Ecosystems: From the Coastline to the Open Sea.

Nearly all aspects of daily life involve plastics, and the production of plastics has increased significantly in recent decades. Plastic is widely used since it is versatile, light, durable, and "cheap." These same qualities allow it to travel over long distances across oceans and seas if we do not manage it carefully. Consequently, plastic is becoming an ever-increasing problem due to its low production cost, potential toxicity, and universal presence. Plastics are persistent materials, which tend to accumulate in the marine environment from coasts to the open sea.

Microplastics (MPs) are defined as all plastic particles <5 mm and come from two sources: (1) primary MPs, which include industrial abrasives, exfoliants, cosmetics, and preproduction plastic pellets; (2) secondary MPs, which come from the fragmentation of plastics. MPs can contain additives such as UV-stabilizers, colorings, flame retardants, and plasticizers, which are transported by the plastic pieces and are susceptible to being taken up by living organisms. These plastics can accumulate persistent organic pollutants (POPs) from the environment and facilitate their transfer into marine food webs. They can also be a vector for invasive species and harmful pathogens.

Although impacts of MPs in marine ecosystems have been reported in the literature since the 70s, many questions remain open. MICRO 2016 provided an opportunity to share available knowledge, fill in gaps, identify new questions and research needs, and develop commitments to operationalize solutions.

Thanks to everyone who submitted abstracts; thanks to everyone who participated in the conference.

The high quality of the scientific material and the engagement of the Organizing Board and Scientific Committee have allowed us to include the work shared at MICRO 2016 in these proceedings. We hope you find them useful, and we wish you a constructive road to MICRO 2018.

Organizing Board and Scientific Committee:

Reserva de la Biosfera de Lanzarote, Cabildo de Lanzarote: **Ana Carrasco and Aquilino Miguelez.**

Observatoire de Versailles SQY: **Juan Baztan, Mateo Cordier and Jean-Paul Vanderlinden.**

Plymouth University: **Sabine Pahl and Richard Thompson.**

Cornell University: **Bethany Jorgensen.**

CNRS: **Thierry Huck, Ika Paul Pont and Philippe Soudant.**

Universitat Politècnica de Catalunya: **Agustín Sánchez-Arcilla.**

Instituto Español de Oceanografía: **Jesús Gago and Luis Valdés.**

ULPGC: **Ana Liria, May Gomez, Alicia Herrera and Ted Packard.**

CSIC: **Elisabetta Broglio and Joaquim Garrabou.**

UNAM: **Ana Carolina Ruiz.**

IFREMER: **François Galgani and Arnaud Huvet.**

Université de Moncton: **Omer Chouinard and Céline Surette.**

Universidad de Cádiz: **Andrés Cozar and Abelardo Gómez Parra.**

Instituut voor Landbouw-en Visserijonderzoek: **Lisa Devrieses and Johan Robbens.**

National Institute of Chemistry: **Andrej Kržan.**

Université Libre de Bruxelles: **Didier Peeters.**

Alfred Wegener Institute: **Melanie Bergmann, Gunnar Gerdts and Lars Gutow.**

The University of Maine: **Jeffrey Runge.**

NUI Galway: **Amy Lusher.**

Université du Maine: **Fabienne Lagarde.**

Universidade Federal de Pernambuco: **Monica Ferreira da Costa and Juliana A. Ivar do Sul.**

Università degli Studi di Siena: **Maria Cristina Fossi.**

SINTEF: **Andy Booth.**

Latvijas Universitate: **Raimonds Ernsteins.**

National Oceanic and Atmospheric Administration: **Carlie Herring and Nancy Wallace.**

UNESCO Intergovernmental Oceanographic Commission: **Henrik Enevoldsen.**

UNESCO MaB: **Miguel Clüsener-Godt.**

Cabildo de Lanzarote

Oficina de la
Reserva de Biosfera

UNIVERSITAT POLITÈCNICA
DE CATALUNYA
BARCELONATECH

ALFRED-WEGENER INSTITUT
HELMHOLTZ-ZENTRUM FÜR POLAR-
UND MEERESFORSCHUNG

UNIVERSITÀ
DI SIENA
1240

UNIVERSITÉ DE
VERSAILLES
ST-QUENTIN-EN-YVELINES

UNIVERSITÉ PARIS-SACLAY

Marine Sciences For Society

UNIVERSITÉ DE MONCTON

Ifremer

UNIVERSIDADE
FEDERAL
DE PERNAMBUCO

European
Commission

SEA
CHANGE
WITH
PLYMOUTH
UNIVERSITY

United Nations
Educational, Scientific and
Cultural Organization

Man and
the Biosphere
Programme

CORNELL UNIVERSITY · FOUNDED A.D. 1865

INSTITUTO ESPAÑOL DE OCEANOGRAFÍA

NUI Galway
OÉ Gaillimh

UNIVERSITÉ
LIBRE
DE BRUXELLES

ULB

Université du Maine

cnrs

UCA

Universidad
de Cádiz

UNIVERSIDAD DE LAS PALMAS
DE GRAN CANARIA

SINTEF

Instituto de Ciencias
del Mar y Limnología
Universidad Nacional Autónoma de México

ICM
CSIC

Institute
of Marine
Sciences

THE UNIVERSITY OF
MAINE

fuem

ILVO

Institute for Agricultural
and Fisheries Research

LATVIJAS
UNIVERSITĀTE
ANNO 1919
U N I V E R S I T Y   O F   L A T V I A

NOAA

NATIONAL OCEANIC AND ATMOSPHERIC ADMINISTRATION
U.S. DEPARTMENT OF COMMERCE

United Nations
Educational, Scientific and
Cultural Organization

Intergovernmental
Oceanographic
Commission

Hosted by:

UNIVERSITÉ DE
VERSAILLES

Cabildo de Lanzarote

Oficina de la
Reserva de Biosfera

Marine Sciences For Society

Under the patronage of UNESCO

CONFERENCE PROGRAMME

May 25th 2016

8h30-9h30 Registration

9h30-10h Welcome and Official Opening

10h-11h20 Session I—Chair persons: Arnaud Huvet and Mateo Cordier.

Ia

Plastic and restricted heavy metals. *A. van Oyen et al.*

Ib

Citizen Research for Ocean Conservation. *M. Boertien et al.*

Ic

Microplastics in Sewage Sludge: Effects of Treatment. *A.-M. Mahon.*

Id

Sources and fate of microplastics in Swiss surface waters. *F. Faure et al.*

Ie

Microplastics in a UK sewage treatment plant. *T. Maes et al.*

11h40-13h Session II—Chair persons:Miguel Clüsener-Godt and Jesús Gago.

IIa

Fates of plastic pollution in a major urban river: persistence and bacterial colonisation of oil-based plastics and bioplastics in the Yarra River, Melbourne, Australia. *T. Gundry et al.*

IIb

Presence and abundance of microplastics in sediments of tributaries of the River Thames, UK. *A. Horton et al.*

IIc

Microplastics in different compartments of the urban water cycle: from the sources to the rivers. *R. Dris et al.*

IId

Validation of a density separation technique for the recovery of microplastics and its use on marine & freshwater sediments. *B. Quinn et al.*

IIe

Microplastics in Singapore's coastal mangrove ecosystems. *N.-H.-M. Nor et al.*

IIf

A quantitative analysis of microplastic pollution along the south-eastern coastline of South Africa. *H. Nel et al.*

15h-16h10 Session III—Chair persons: May Gomez and Alicia Herrera.

IIIa

Plastic pollutants within the marine environment of Durban, KwaZulu-Natal, South Africa. *T. Naidoo et al.*

IIIb

Reading between the grains − Microplastics in intertidal beach sediments of a World Heritage Area: Cleveland Bay (QLD), Australia. *R. Schoeneich-Argent et al.*

IIIc

Patterns of plastic pollution in offshore, nearshore and estuarine waters of Perth, Western Australia. *S. Hajbane et al.*

IIId

Microplastic Distribution and Composition in the Israeli Mediterranean Coastal Waters. *N. van der Hal et al.*

IIIe

Occurrence of microplastics in the South Eastern Black Sea. *U. Aytan et al.*

IIIf

Microplastics migrations in sea coastal zone: Baltic amber as an example. *I. Chubarenko.*

IIIg

Simultaneous trace analysis of nine common plastics in environmental samples via pyrolysis gas chromatography mass spectrometry (Py-GCMS). *M. Fischer et al.*

16h30-17h30 Session IV—Chair persons: Maria Cristina Fossi and Ana Liria.

IVa

Extensive review on the presence of microplastics and nanoplastics in seafood: data gaps and recommendations for future risk assessment for human health. *K. Mackay et al.*

IVb

State of knowledge on human health implications on consumption of aquatic organisms containing microplastics. *E. Garrido Gamarro et al.*

IVc

Uptake and Ecotoxicity of microplastic particles (polystyrene) by *Daphnia magna. R. Aljaibachi et al.*

IVd

Effects of PVC and Nylon microplastics on survival and reproduction of the small terrestrial earthworm *Enchytreus crypticus. A. Walton.*

IVe

Microplastics: who is at risk? *R. Saborowski et al.*

IVf

Engagement options for the implementation of the European Atlantic Plan. *F. Cardona.*

17h30-19h Posters and side event "Education"

19h-20h30 Session V—Chair persons: Sabine Pahl and Jean-Paul Vanderlinden.

Va

Understanding Microplastic Distribution: A Global Citizen Monitoring Effort. *A. Barrows.*

Vb

Monitoring of plastic pollution in the North Sea. *E. Leemans.*

Vc

Voluntary beach cleanups at Famara Beach, Lanzarote. *N.A. Ruckstuhl et al.*

Ve

The Wider Benefits of Cleaning Up Marine Plastic: Two Psychological Studies Examining the Direct Impacts of Beach Cleans and Fishing for Litter on the Volunteers. *K. Wyles et al.*

Vf

Agüita con el Plástico: Society as part of the solution of plastic pollution. *J.-C. Jiménez et al.*

Vg

No Plastic campaign makes a difference in Island of Principe Biosphere Reserve. *M. Clüsener-Godt et al.*

Vh

Logistics of Coastline Plastic Cleanup and Recycling: A survey and framework. *N. Brahimi et al.*

Vi

Tackling microplastics on land: citizen observatories of anthropogenic litter dynamics within the MSCA POSEIDOMM project. *L. Galgani et al.*

Vj

Environmental Science Education – Methodologies to promote Ocean Literacy. *F. Silva et al.*

Discussion and conclusions

May 26th 2016

9h-10h30 Session VI—Chair persons: Lars Gutow and Fabienne Lagarde.

VIa

Microplastics, convergence areas and fin whales in the northwestern Mediterranean Sea. *M.-C. Fossi et al.*

VIb

Microplastics in marine meso-herbivores. *L. Gutow et al.*

VIc

Investigating the Presence and Effects of Microplastics in Sea Turtles. *E. Duncan et al.*

VId

Microplastics presence on sea turtles. *P. Ostiategui-Francia et al.*

VIe

Factors determining the composition of plastics from the South Pacific Ocean – Are seabirds playing a selective role? *V. Hidalgo-Ruz et al.*

VIf

Micro- and macro-plastics associated with marine mammals stranded in Ireland: Recent findings and a review of historical knowledge. *A. Lusher et al.*

VIg

Primary (ingestion) and secondary (inhalation) uptake of microplastic in the crab Carcinus maenas, and its biological ef-fects. *A. Watts et al.*

VIh

Plastic in Atlantic cod (Gadus morhua) from the Norwegian coast. *D. Pettersen Eidsvoll et al.*

VIi

Extraction and characterization of microplastics in marine organisms sampled at Giglio Island after the removal of the Costa Concordia wreck. *C. Giacomo Avio et al.*

11h-13h Session VII—Chair persons: Gunnar Gerdts and Johan Robbens.

VIIa

Floating plastic marine debris in the Balearic Islands: Ibiza case study. *M. Compa et al.*

VIIb

Enzymes – essential catalysts in biodegradation of plastics. *G. Guebitz et al.*

VIIc

Studies of microplastics from the Irish marine environment. *A. Lusher.*

VIId

Deposition of microplastics in marine sediments from the Irish continental shelf. *J. Martin et al.*

VIIe

Where go the plastics? And whence do they come?

From diagnosis to participatory community-based observatory network. *J. Baztan et al.*

VIIf

Distribution and composition of microplastics in Scotland's seas. *M. Russell et al.*

VIIg

Marine litter accumulation in the Azorean Archipelago: Azorlit preliminary data. *J. Frias et al.*

VIIh

Microplastics in the Adriatic — Results from the DeFishGear project. *A. Palatinus et al.*

VIIi

Floating microplastics in Mediterranean surface waters. *G. Suaria et al.*

VIIj

Implementation of the Spanish Monitoring Program of Microplastics on Beaches within the Marine Strategy Framework Directive. First phase. *J. Buceta et al.*

VIIk

Operational forecasting as a tool for managing pollutant dispersion and recovery. *A. Sánchez-Arcilla et al.*

15h-16h10 Session VIII—Chair persons: Melanie Bergmann and Juan Baztan.

VIIIa

What do we know about the ecological impacts of microplastic debris? *C. Rochman et al.*

VIIIb

Qualitative and quantitative investigations of microplastics in pelagic and demersal fish species of North and Baltic Seas using pyrolysis-GCMS. *B. Scholz-Böttcher et al.*

VIIIc

Microplastics extraction method from small fishes, on the road to the standard monitoring approach. *S. Budimir et al.*

VIIId

Microplastic effects in Mullus surmuletus: Ingestion and induction of detoxification systems. *C. Alomar et al.*

VIIIe

VIIIe1 Assessment of microplastics present in mussels collected from the Scottish coast. *A. I Catarino et al.; VIIIe2* Bioavailability of co-contaminants sorbed to microplastics in the blue mussel *Mytilus edulist. A. I. Catarino et al.*

VIIIf

Exploring the effects of microplastics on the hepatopancreas transcriptome of *Mytilus galloprovincialis. M. Milan et al.*

16h30-17h30 Session IX—Chair persons: Elisabetta Broglio and Ted Packard.

IXa

The Characterisation, Quantity and Sorptive Properties of Microplastics from Cosmetics. *I. Napper et al.*

IXb

Beach sweep initiatives on the Acadian Coastline in Atlantic Canada. *O. Chouinard et al.*

IXc

A social-ecological approach to the problem of floating plastics in the Mediterranean Sea. *L. Ruiz Orejon et al.*

IXd

News splash? A preliminary review of microplastics in the news. *B. Jorgensen et al.*

IXe

Informing policy makers about state of knowledge and gaps on microplastics in the marine environment. *T. Bahri.*

IXf

Microplastics in cosmetics: Exploring perceptions of environmentalists, beauticians and students. *S. Pahl et al.*

17h30-19h Posters

19h-20h30 Public Keynote: "Dialogue between natural and social sciences"

By François Galgani and Jean-Paul Vanderlinden.

The Poster Awards will be announced at 20h27...

May 27th 2016

9h-10h30 Session X—Chair persons: Céline Surette and Quino Miguelez.

Xa

Plastics and Zooplankton: What do we know? *P. Lindeque et al.*

Xb

Microtrophic Project. *A. Herrera et al.*

Xc

Source to sink: Microplastics in benthic fauna in a gradient from discharge points to deep basins in an urban model fjord. *M. Haave et al.*

Xd

Uptake and toxicity of methylmethacrylate-based nanoplastic particles in aquatic organisms. *A. Booth et al.*

Xe

On the potential role of phytoplankton aggregates in microplastic sedimentation. *M. Long et al.*

Xf

Hitchhiking microorganisms on microplastics in the Baltic Sea. *S. Oberbeckmann et al.*

Xg

Microplastic Prey? – An assay to investigate microplastic uptake by heterotrophic nanoflagellates. *A. Madita Wieczorek et al.*

Xh

Microplastics in seafood: identifying a protocol for their extraction and characterization. *G. Duflos*

11h-13h Session XI—Chair persons: Andrej Kržan and Bethany Jorgensen.

XIa

Vast quantities of microplastics in Arctic sea ice – a prime temporary sink for plastic litter and a medium of transport. *M. Bergmann et al.*

XIb

Microplastics in the Bay of Brest (Brittany, France): composition, abundance and spatial distribution. *L. Frère et al.*

XIc

Plastic litters: a new habitat for marine microbial communities. *C. Dussud et al.*

XId

The effects of microplastic on freshwater *Hydra attenuatta* morphology & feeding. *F. Murphy.*

XIe

Evidence of microplastic ingestion in elasmobranchs in the western Mediterranean Sea. *S. Deudero et al.*

XIf

Do microplastic particles impair the performance of marine deposit and filter feeding invertebrates? Results from a globally replicated study. *M. Lenz et al.*

XIg

Occurrence of potential microplastics in commercial fish from an estuarine environment: Preliminary results. *F. Bessa et al.*

XIh

Improvements and needs of microplastics analytical control at open sea − opportunities for monitoring at Canary Islands. *D. Vega-Moreno et al.*

15h-16h10 Session XII—Chair persons: Andy Booth and Raimonds Ernsteins.

XIIa

A novel method for preparing microplastic fibers. *M. Cole.*

XIIb

A new approach in processing freshwater suspended particulate matter for separating microplastics. *S. Hatzky.*

XIIc

Microplastics − microalgae: an interaction dependent on polymer type. *F. Lagarde et al.*

XIId

Nearshore circulation in the Confital Bay: Implications on marine debris transport and deposition at Las Canteras Beach. *L. Mcknight Morales et al.*

XIIe

New approaches of the extraction and identification of microplastics from marine sediment. *M. Kedzierski et al.*

16h30-17h30 Session XIII—Chair persons: Amy Lusher and Omer Chouinard.

XIIIa

Using physical and chemical characteristics of floating microplastics to investigate their weathering history. *K. Law et al.*

XIIIb

The size spectrum as tool for analyzing marine plastic pollution. *E. Marti et al.*

XIIIc

Automated analysis of μFTIR imaging data for microplastic samples. *S. Primpke et al.*

XIIId

Solar radiation induced degradation of common plastics under marine exposure conditions. *A. Andrady et al.*

XIIIe

Using the FlowCam to validate an enzymatic digestion protocol applied to assess the occurrence of microplastics in the Southern North Sea. *C. Lorenz et al.*

XIIIf

DNA from the 'Plastisphere': an extraction protocol for ocean microplastics. *P. Debeljak et al.*

XIIIg

Quality assurance in microplastic detection. *C. Wesch et al.*

17h30-19h Posters and side event "Sharing Responsibilities"

19h Conclusions: Lanzarote's Declaration

Posters

Dear participants, All posters can be hung up starting the morning of May 25th at 9h after registration. They must be removed by 19h on Friday the 27th.

-Note: Posters are listed in alphabetical order of 1st author, please keep the poster number in mind for the posters contest-.

The Poster Awards will be announced on the 26th at 20h27'

during the Public Keynote: "Dialogue between natural and social sciences".

Are the densities of microplastics altered following interactions with *Elminius modestus* and sediment particles? *R. Adams et al.*, poster number **102175**

Macro- and micro-plastic in seafloor habitats around Mallorca. *C. Alomar et al.*, poster number **102179**

Evaluation of microplastics in Jurujuba Cove, Niterói. *F. Araujo et al.*, poster number **94158**

Presence, distribution and characterization of microplastics in commercial organisms from Adriatic Sea. *C-G Avio et al.*, poster number **101916**

The origin and fate of microplastics in saltmarshes. *H. Ball et al.*, poster number **101818**

Effects of microplastics and mercury, alone and in mixture, on the European sea bass (*Dicentrarchus labrax*): swimming performance and sub-individual biomarkers. *L-G Barboza et al.*, poster number **101915**

DNA damage evaluation of Polyethylene microspheres in *Daphnia magna*. *A. Berber et al.*, poster number **102315**

LITTERBASE – An online portal for marine litter and microplastics and their implications for marine life. *M. Bergmann et al.*, poster number **102482**

PLASTOX: Direct and indirect ecotoxicological impacts of microplastics on marine organisms. *A. Booth et al.*, poster number **95410**

Community-based Observatories tackling MICROPLASTIC: Spanish schools pilot project based on seawatchers.org. *E. Broglio et al.*, poster number **101794**

Tackling marine litter: Awareness and outreach in the Azores. *R. Carriço et al.*, poster number **103441**

Prevalence of microplastics in the marine waters of Qatar's Exclusive Economic Zone (EEZ). *A. Castillo et al.*, poster number **100223**

Microplastics contamination in three planktivorous and commercial fish species. *F. Collard et al.*, poster number **100959**

Spatial variation of microplastic ingestion in *Boops boops* in the Western Mediterranean Sea. *M. Compa et al.*, poster number **102142**

Source, transfer and fate of microplastics in the nothwestern Mediterranean Sea: a holistic approach. *M. Constant et al.*, poster number **102220**

Microplastics in the final ocean frontier. *W. Courtene-Jones et al.*, poster number **99166**

Microplastic abundance and distribution in the intertidal and subtidal marine environment around a major urban park in Vancouver, Canada. *A. Diaz et al.*, poster number **102008**

Microplastic ingestion by decapod larvae. *K. Reilly et al.*, poster number **101119**

May polystyrene microparticles affect mortality and swimming behaviour of marine planktonic invertebrates? *Ch. Gambardella et al.*, poster number **101801**

Improvement of microplastic extraction method in organic material rich samples. *P. Garrido Amador et al.*, poster number **102239**

Defining the Baselines and standards for Microplastics Analyses in European Waters (JPI-O BASEMAN). *G. Gerdts*, poster number **100987**

Effects of long term exposure with contaminated and clean micro plastics on *Mytilus edulis*. *T. Hamm*, poster number **102416**

The City of Kuopio and Lake Kallavesi in the Finnish Lake District a 'living laboratory' for the microplastic pollution research in freshwater lakes. *S. Hartikainen et al.*, poster number **102292**

Analysis and quantification of microplastics in the stomachs of common dolphin (*Delphinus delphis*) stranded on the Galician coasts (NW Spain). *A. Hernandez-Gonzalez et al.*, poster number **99376 and 101623**

First report of microplastics in bycaught pinnipeds. *G. Hernandez-Milian et al.*, poster number **102200**

Preliminary results of Microtrophic Project. *A. Herrera et al.*, poster number **101570**

The Contribution of Citizen Scientists to the Monitoring of Marine Litter. *V. Hidalgo-Ruz et al.*, poster number **103474**

Experimenting on settling velocity of cylindrical microplastic particles. *I. Isachenko et al.*, poster number **101528**

WEATHER-MIC – How microplastic weathering changes its transport, fate and toxicity in the marine environment. *A. Jahnke et al.*, poster number **91388**

Uptake of textile polyethylene terephthalate microplastic fibres by freshwater crustacean *Daphnia magna*. *A. Jemec et al.*, poster number **103222**

Microplastic distribution on two northwestern Mediterranean beaches. *P. Kerhervé et al.*, poster number **102341**

Effects of microplastics on digestive enzymes in the marine isopod Idotea emarginata. *Š. Korez et al.*, poster number **102408**

Sinking behaviour of microplastics. *N. Kowalski et al.*, poster number **100878**

Characteristics of Plastic in stomachs of Northern Fulmars (*Fulmarus glacialis*). *S. Kühn et al.*, poster number **100942**

Microplastic sampling in the Mediterranean Sea. *J. Larsen et al.*, poster number **101871**

Pilot study on microlitter in the surface waters of the Gulf of Finland, Baltic Sea. *K. Lind et al.*, poster number **101411**

Persistent Organic Pollutants adsorbed on microplastics from the North Atlantic gyre. *M. Martignac et al.*, poster number **102442**

First evidence of microplastics in the ballast water of commercial ships. *M. Matiddi et al.*, poster number **101548**

Microlitter abundance in the Italian Minor Islands, Central Mediterranean Sea. *M. Matiddi et al.*, poster number **101615**

In search of the plastic accumulation regions: fine-tuning ocean surface transport models. *R. McAdam et al.*, poster number **101799**

Preliminary assessment of the microplastic presence in the Gulf of Genoa (Italy, Ligurian Sea, Northwestern Mediterranean Sea). *S. Morgana et al.*, poster number **101802**

Toxicity assessment of pollutants sorbed on microplastics using various bioassays on two fish cell lines. *B. Morin et al.*, poster number **102337**

Fate of microplastics in soft marine sediments. *P. Näkki et al.*, poster number **102107**

Microplastic as a vector of chemicals to fin whale and basking shark in the Mediterranean Sea: A model-supported analysis of available data. *C. Panti et al.*, poster number **101611**

Methodological prerequisites for toxicity testing of microplastics using marine organisms. *J.-W. Park et al.*, poster number **102067**

POPs adsorbed on plastic pellets collected in the Adriatic region. *M. Pflieger et al.*, poster number **102271**

The EPHEMARE Project: Ecotoxicological Effects of Microplastics in Marine Ecosystems. *F. Regoli*, poster number **101391**

Suspended micro-sized PVC particles impair the performance and decrease survival in the Asian green mussel *Perna viridis*. *S. Rist*, poster number **101884**

Priority pollutants in microplastics from beaches in Gran Canaria Island. *M. Rodrigo et al.*, poster number **88302**

Plastic debris in Mediterranean seabirds. *A. Rodríguez et al.*, poster number **110743**

Monitoring plastic ingestion in selected Azorean marine fauna: Azorlit preliminary data. *Y. Rodriguez et al.*, poster number **101934**

Catching a glimpse of the lack of harmonization regarding techniques of extraction of microplastics in marine sediments. *E. Rojo-Nieto et al.*, poster number **96241**

Floating plastics in the sea: People's perception in the Majorca island (Spain). *L. Ruiz Orejon et al.*, poster number **101076**

Plastics in the Mediterranean Sea surface: From regional to local scale. *L. Ruiz Orejon et al.*, poster number **101081**

Analysis of organic pollutants in micro-plastics. *S. Santana-Viera et al.*, poster number **101858**

From the sea to the dining table and back to the environment: micro-litter load of common salts. *O. Setälä et al.*, poster number **102227**

Sandy beaches microplastics of the Crimea Black Sea Coast. *E. Sibirtsova et al.*, poster number **100994**

Plastic prey; are fish post-larval stages ingesting plastic in their natural environment? *M. Steer et al.*, poster number **101113**

Microplastic ingestion by planktivorous fishes in the Canary Current. *A. Štindlová et al.*, poster number **102530**

First quantification of microplastic in Norwegian fjords through non disruptive ad-hoc sampling. *M. Svendsen Nerheim et al.*, poster number **103400**

Abundance of microplastics and adhered contaminants in the North Atlantic Ocean. *K. Syberg et al.*, poster number **101866**

Bioplastic and microbes. *V. Turk et al.*, poster number **102595**

Marine litter monitoring for coastal management indicator system development: citizen science and collaboration communication approach. *J. Ulme et al.*, poster number **110276**

Types and concentration of microplastics found on remote island beaches during the Race for Water Odyssey. *C. Levasseur et al.*, poster number **99434**

A throwaway society: Is science stuck with single use plastic? What can we do about it? *A. Watts et al.*, poster number **100868**

Linking education and science to increase awareness of marine plastic litter – Distribution of plastic waste on beaches of the German Bight. *A. Wichels et al.*, poster number **98897**

Are smaller microplastics underestimated? Comparing anthropogenic debris collected with different mesh sizes. *A. Wilson Mcneal et al.*, poster number **100984**

Precipitation/Flotation Effect of Coagulant to microplastics in water. *M. Yurtsever et al.*, poster number **105617**

Personal care and cosmetics products (PCCPs): Is it cleaning or pollution? *M. Yurtsever et al.*, poster number **105618**

Detection of microplastics with stimulated Raman scattering (SRS) microscopy. *L. Zada et al.*, poster number **102003**

PART *I*

Abstracts From
Oral Presentations

Plastic and Restricted Heavy Metals

A. van Oyen[1], S. Kühn[2], J.A. van Franeker[2], M. Ortlieb[3]
and M. Egelkraut-Holtus[3]

[1]CARAT GmbH, Bocholt, Germany [2]IMARES, Den Helder, The Netherlands
[3]Shimadzu Europa GmbH, Duisburg, Germany

Plastic has become an integral part of our daily life and its use is increasing. In 2014 the worldwide production has reached an all time high of 311 million tons. Single use-packaging, mainly food, accounts for almost 40% of the total production in the EU. Modern plastics for food packaging have to be safe (EU Commission Regulation, 2011), but is this always the case? In PET, used for instance in bottles and tea bags, a toxic leftover of the catalyst Sb_2O_3 can be found. These leftovers could migrate from plastic into the beverage. Could the inheritance of the past contaminate the future? Carbon-based plastics are thermodynamically metastable and will degrade over time. Heavy metals are firmly bound in plastic but degradation could accelerate migration of heavy metals. In the past the Life Cycle Assessment was linear: after usage plastic became waste and ended mainly as landfill or thermal recycling. Under consumer and political pressure the EU indicated that it has to become a circular economy. Plastics of durable applications, like cars, electronics, and crates, make recycling more difficult. During their functional life new regulations have been introduced. In the EU several regulations have been developed over the past decades, the recycled raw materials of recyclates could be contaminated with the inheritance of the past. Nowadays plastic is found littering the environment in large quantities. The ingestion of plastic by seabirds is best known and monitored, but the phenomenon of ingesting plastics is widespread among all marine biota (Kühn et al., 2015). New investigations prove that plastics loaded with heavy metals are found in the environment, which when ingested by wildlife may pose specific additional toxicity risks which we investigate in the JPI Oceans PLASTOX project.

REFERENCES

EU Commision Regulation, (EU) No 10/2011 of 14 January 2011 on plastic materials and articles intended to come into contact with food. http://eur-lex.europa.eu/legal-content/EN/ALL/?uri = CELEX%3A32011R0010.

Kühn, S., Bravo Rebolledo, E.L., Van Franeker, J.A., 2015. Deleterious effects of litter on marine life. In: Bergmann, M., Gutow, L., Klages, M. (Eds.), Marine Anthropogenic Litter. Springer, Berlin, pp. 75–116, http://edepot.wur.nl/344861 (includes supplement).

MICRO 2016. Fate and Impact of Microplastics in Marine Ecosystems.

Citizen Research for Ocean Conservation

M. Boertien and E. Butter
Ocean Conservation

For further information on this study, please contact the authors.

Microplastics in Sewage Sludge: Effects of Treatment

A.M. Mahon[1], B. O' Connell[1], M.G. Healy[2], I. O' Connor[1], R. Officer[1], R. Nash[1] and L. Morrison[2]

[1]Galway-Mayo Institute of Technology, Galway, Ireland [2]National University of Ireland, Galway, Ireland

Urban Waste Water Treatment Plants (UWWTPs) are receptors for the cumulative loading of microplastics derived from industry, landfill, domestic waste water, and storm water. The partitioning of microplastics through the settlement processes of waste water treatment results in the majority becoming entrained in the sewage sludge. In the EU, sewage sludge is dealt with in a wide variety of ways, including use as an agricultural fertilizer for which it is required that human health risks be minimized through heat, chemical, or biological treatment. This study characterized microplastics in sludge samples in 7 UWWTPs in Ireland using anaerobic digestion (AD), thermal drying (TD), and lime stabilization (LS) treatment processes. Abundances ranged from 4196 to 15,385 particles/kg (dry weight). Results of a general linear mixed model showed significantly higher abundances of microplastics in smaller size classes in LS samples, suggesting that the treatment process of LS sheer microplastic particles. In contrast, lower abundances of microplastics found in the AD samples suggests that this process may reduce microplastic abundances. A range of polymer types were identified using Fourier Transform Infrared Spectroscopy and surface morphologies examined using Scanning Electron Microscopy showed characteristics of melting and blistering of TD microplastics and shredding and flaking of LS mircoplastics. This study highlights the potential for sewage sludge treatment processes to increase or reduce the risk of microplastic pollution prior to land spreading and may have implications for legislation governing land-spreading of biosolids.

Sources and Fate of Microplastics in Swiss Surface Waters

F. Faure, O. Pompini and L.F. de Alencastro
Central Environmental Laboratory, EPFL, Lausanne, Switzerland

While plastic pollution in marine environments is getting better known and is the subject of a continuously growing number of publications, freshwaters still receive too little attention. Some studies worldwide focused on the nature and concentrations of microplastics in freshwater bodies, but data is so far very limited and fragmentary. Microplastics were studied in Swiss surface waters since 2012, first in order to show the reality of this contamination, and later to better understand its nature and extent. Microplastics were found in significant concentrations in all the sampled lakes and rivers, both on the water surface, beach sediments, and benthic sediments, as well as in organisms (fishes, birds, and zebra mussels). They were shown to contain potentially toxic additives, as well as to adsorb hydrophobic contaminants (Faure et al., 2015). Some pathways into the water bodies have then been studied including Waste Water Treatment Plant (WWTPs) urban runoff waters, soils or storm overflows. Potential sources and pathways of (micro-) plastics in the Lake Geneva watershed are now the focus, on the basis of the fractions of plastics that enter the environment at all stages of their life cycle depending on their use, and on the plastic types and quantities that can be found in the different environmental compartments. This dual approach of material flow modeling and a strong integration of measured environmental data is the first step towards producing a mass balance of microplastics in the Lake Geneva watershed.

REFERENCE

Faure, F., Demars, C., Wieser, O., Kunz, M., de Alencastro, L.F., 2015. Plastic pollution in Swiss surface waters: nature and concentrations, interaction with pollutants. Environ. Chem. 12 (5), 582–591.

Microplastics in a UK Sewage Treatment Plant

T. Maes[1] and H. Leslie[2]

[1]CEFAS, Centre for Environment, Fisheries, Aquaculture and Science, Lowestoft, UK [2]VU University Amsterdam, Amsterdam, The Netherlands

For further information on this study, please contact the authors.

Fates of Plastic Pollution in a Major Urban River: Persistence and Bacterial Colonization of Oil-based Plastics and Bioplastics in the Yarra River, Melbourne, Australia

T. Gundry, B. Clarke and A.M. Osborn

RMIT University, Bundoora, VIC Australia

For further information on this study, please contact the authors.

Presence and Abundance of Microplastics in Sediments of Tributaries of the River Thames, UK

A.A. Horton, C. Svendsen, R.J. Williams, D.J. Spurgeon and E. Lahive

Centre for Ecology and Hydrology, Wallingford, United Kingdom

Microplastics are becoming widely recognized as a global environmental contaminant but with the exception of estuaries, there are currently no published data on microplastic presence within any UK freshwater bodies. The aim of this study was to identify and quantify large microplastic particles (1–4 mm) from a range of tributaries in the Thames River Basin, UK. Four sites were chosen based on potential for contamination, covering a range of population densities and sewage effluent inputs. Microplastics of two size fractions (1–2 mm and 2–4 mm) were extracted from sediments using a stepwise approach to include sieving, sorting by eye, flotation, and identification using Raman spectroscopy.

The highest number of microplastic particles was found at the site with the second-highest effluent input, with plastic particles also found at lesser-impacted rural sites. At the site with highest abundance of microplastic particles (average 66 particles/100 g) the dominant particle type was fragments, some of which could be directly traced to road-marking paints or road surface coatings, implying that effluent is not the dominant input here. At all other sites fibers were the dominant particle type. Raman spectroscopy identified polymers including polyarylsulphone, polyethylene, polystyrene, polypropylene, and polyester, in addition to polymer-associated dyes. This study is the first to identify and describe the presence of microplastics in a UK freshwater system (Thames basin sediments) at both urban and rural locations. Furthermore road-derived particles, a previously undescribed source of environmental microplastics, were identified as a source. These results highlight rivers as being a potential transport pathway for transfer of microplastic particles from terrestrial sources to the oceans.

Microplastics in Different Compartments of the Urban Water Cycle: From the Sources to the Rivers

R. Dris[1], J. Gasperi[1], B. Bounoua[1], V. Rocher[2] and B. Tassin[1]
[1]University of Paris-Est, Créteil, France [2]SIAAP (syndicat interdépartemental pour l'assainissement de l'agglomération parisienne), Colombes, France

If the occurrence of microplastics, particles smaller than 5 mm, in both marine and continental environments have been studied, their inputs are still very poorly identified and few works focused on their sources. Moreover only few studies are dealing with the occurrence of microplastics on rivers. This work aims at assessing the microplastic contamination from different compartments of the urban water cycle including surface water. For surface waters, a sub-objective is to estimate the spatiotemporal variability of microplastics in a river. Greywater and wastewater were first studied. In washing machine effluents, concentrations between 8,850,000 and 18,700,000 particles/ m^3 were encountered, confirming the large contribution of the clothes as a source of fibers. Wastewater treatment plant influents were also analyzed exhibiting high levels of fibrous plastics

(260,000–320,000 particles/m^3) while the concentrations in the effluents are in the 14,000–50,000 particles/m^3 range. A removal rate
between 83 and 95% has been estimated. Throughout the monitoring,
an average atmospheric fallout of 110 × 96 particles/m^2/day
(mean × SD) was encountered in the urban site against 53 × 38 parti-
cles/m^2/day on the suburban site. A significant difference between both
sites was found. The Marne and Seine Rivers were also considered.
Globally, levels vary in the 1.04–441.38 particles/m^3 range in surface
water. During short-term temporal variability tests, concentrations ranged between 38.2 and 101.6 particles/m^3 in the first campaign (variability about 45%, $n = 3$) and between 18.7 and 38.6 particles/m^3 during the
seconds (26%, $n = 3$). A lateral variability ($n = 9$) of 53% is found. The
vertical variability shows a coefficient of variation of 21% ($n = 9$) which
is more than two times lower than the lateral variability. This result is
not surprising knowing that in river conditions, a constant water mixing
is induced by the currents and reduces the chances of stratification.

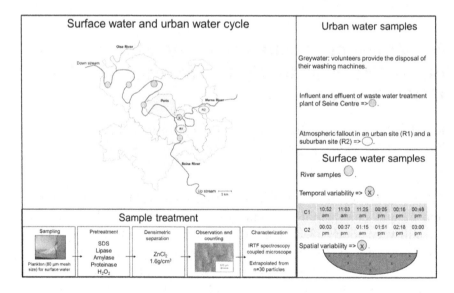

Synthetic figure of the sites, considered compartments and the sampling and treatment approaches.

Validation of a Density Separation Technique for the Recovery of Microplastics and Its Use on Marine Freshwater Sediments

B. Quinn, F. Murphy and C. Ewins
University of the West of Scotland, Paisley, Scotland

Currently there is no standardized method for the collection or separation of microplastics from sediment samples, hindering the generation of data on their presence and potential environmental impact. Density separation using various brine solutions has been a popular method for extracting microplastics from sediment. The aims of this work were (1) to validate a density separation method using a new brine solution of zinc bromide in comparison to other solutions for separating microplastics from sediments and (2) to apply this method to marine and freshwater sediment samples to isolate and identify environmental microplastics. The efficiency of four brine solutions (sodium chloride, sodium bromide, sodium iodide, and zinc bromide) and water to separate out microplastics from a marine sediment (200–400 μm) spiked with different plastics was tested. The plastics included polyethylene, high density polyethylene, nylon, polyethylene terephthalate, polystyrene, and polyvinyl chloride. From the validation test it is evident that the best recoveries were obtained with both the sodium iodide and the zinc bromide solution. The zinc bromide brine solution was used to test for microplastics in marine and freshwater sediments taken along the firth of Clyde, Scotland. Density separation using zinc bromide brine solution is an effective method for the separation of microplastics from sediments. Although expensive, zinc bromide can be successfully reused indefinitely (once resaturated to 25%) and is a cost effective and efficient method as high recovery rates are achieved after one sample run, as opposed to the three runs needed when using sodium chloride.

Microplastics in Singapore's Coastal Mangrove Ecosystems

N.H.M. Nor[1] and J.P. Obbard[2]
[1]National University of Singapore, Singapore [2]Qatar University, Qatar

The prevalence of microplastics in the marine environment has been extensively reported in the Atlantic and Pacific oceans and their coastal regions. To date, little is known about the microplastics contamination in the marine environment of Southeast Asia which lies within one of the busiest maritime regions of the world. Singapore is a small island nation located in this region and due to intensive shipping activity and coastal development, marine debris is now ubiquitous in its offshore and coastal areas. In this study, the abundance of microplastics was studied in seven intertidal tropical mangrove habitats on Singapore's coastline. Mangrove, which is typified by aerial and stilt roots, is able to trap and accumulate marine debris resulting in long term entrapment and degradation of plastic into microplastic particles over time. Microplastics (1.6 μm−5 mm) were extracted from mangrove sediments via a floatation method, and then counted and categorized according to particle shape and size. Representative microplastics from Berlayar Creek, Sungei Buloh, Pasir Ris, and Lim Chu Kang were isolated for polymer identification using Attenuated Total Reflectance-Fourier Transform Infrared (ATR-FTIR) spectroscopy. Microplastics were identified in all seven habitats ranging from 12.0 to 62.7 particles/kg dry sediment, with the highest concentration at Lim Chu Kang in the northwest of Singapore. Maximum concentrations found in this study were about 3 times higher than the microplastics occurrence along the Belgian coastline and similar to the maximum concentrations found in the UK subtidal and estuarine habitats. Majority of microplastics found were fibrous and less than 20 μm. Four polymer types were identified from the selected microplastics analyzed under ATR-FTIR, including: polyethylene, polypropylene, nylon, and polyvinyl chloride. Higher concentrations occurred at sites where more offshore activities are found and the polymer types suggest that they originated from sea-based activities such as fish farms.

A Quantitative Analysis of Microplastic Pollution Along the South-eastern Coastline of South Africa

H.A. Nel and P.W. Froneman
Rhodes University, Grahamstown, South Africa

The extent of microplastic pollution (<5 mm) in the southern hemisphere, particularly southern Africa, is largely unknown. This study aimed to evaluate microplastic pollution ($>65\,\mu$m) along the southeastern coastline of South Africa, looking at whether bays are characterized by higher microplastic densities than open stretches of coastline in both beach sediment and surf-zone water. Microplastic (mean $\pm$ standard error) densities in the beach sediment ranged between 688.9 ± 348.2 and 3308 ± 1449 particles/m^2, while those in the water column varied between 257.9 ± 53.36 and 1215 ± 276.7 particles/m^3. With few exceptions there were no significant spatial patterns in either the sediment or water column microplastic densities; with little differences in density between bays and the open coast ($p < 0.05$). Additionally, samples were made up entirely of secondary microplastics (fibers and fragments), thus no primary microplastics, such as industrial pellets and microbeads from cosmetics were found during this study possibly due to the study area being situated in a low industrialized region. Fibers were categorized according to color, of which blue/black and red were the predominant forms. South Africa, a developing country runs the risk of rapid population growth followed by a collapse in infrastructure. Beaches along the south eastern coastline of South Africa have often been described as "pristine" and "unspoilt." However, this study, which is the first published article evaluation on the status of microplastic pollution in this area showed beach sediment and water samples are in fact heavily polluted. Especially by microfibers originating from the fragmentation of synthetic items (i.e., garments and carpets). Additionally, these data indicate that the presence of microplastics were not associated with proximity to land-based sources or population density, but rather is governed by water circulation. However, this needs to be investigated further on a national level.

Plastic Pollutants Within the Marine Environment of Durban, KwaZulu-Natal, South Africa

T. Naidoo[1], D. Glassom[1] and A. Smit[2]

[1]University of KwaZulu-Natal (UKZN), Durban, South Africa [2]University of the Western Cape, Bellville, South Africa

Widespread disposal of plastics negatively affects biotic and abiotic components of marine systems. Monitoring plastic concentrations in estuaries is vital in assessing the magnitude of terrestrial inputs to oceanic environments. Data on plastics ≤5 mm in estuaries are scant. This study determined microplastic levels within five estuaries along the Durban coastline, South Africa, and on their intervening beaches. Plastics were isolated from estuarine sediment, beach sediment, and the surface water of each estuary and characterized. Sediment at the Bayhead area of Durban harbor had the highest average plastic concentrations (745.40 ± 129.72 particles per 500 mL). Overall, an attenuating concentration trends away from the city center was found. Fragments composed the largest percent of plastics (59%) found in Bayhead, whereas fibers dominated other estuaries with proportions ranging from 38% of total plastics in the uMgeni estuary to 66% in the Mdloti.

Reading Between the Grains – Microplastics in Intertidal Beach Sediments of a World Heritage Area: Cleveland Bay (QLD), Australia

R. Schoeneich-Argent[1,2,3], A. Negri[2], F. Kroon[2], M. Hamann[1] and L. Van Herwerden[1]

[1]James Cook University (JCU), Townsville, QLD, Australia [2]Australian Institute of Marine Science, Townsville, QLD, Australia [3]Carl von Ossietzky University Oldenburg, Wilhelmshaven, Germany

For further information on this study, please contact the authors.

Patterns of Plastic Pollution in Offshore, Nearshore, and Estuarine Waters of Perth, Western Australia

S. Hajbane and C. Pattiaratchi

The University of Western Australia (CEME), Crawley, WA, Australia

For further information on this study, please contact the authors.

Microplastics in Israeli Mediterranean Coastal Waters

Noam Van Der Hal[1], Lorena M. Rios[2] and Dror L. Angel[1]

Department of Maritime Civilizations, Charney School for Marine Science, University of Haifa, Haifa, Israel The Department of Natural Science, University of Wisconsin - Superior, USA.

Floating microplastics abundances were monitored over a 2-year period (2013–15) along the Israeli Mediterranean coast using a Manta trawl. The microplastics were analyzed for the presence of POPs using soxhlet extraction and GC-MS for identification and quantitation. In addition, shallow (top 10   cm) sediment samples were collected by scuba divers at 10   m depth to explore these for presence of microplastics at the sea-floor. Moreover, microplastic particles were identified by polymer type using FTIR. Microplastics were found in all of the sea surface samples examined, at all sites and in all seasons sampled, with an overall mean abundance of 7.68 particles/m^3 $\pm$ 26.53. On average, abundances of floating microplastics recorded in this study were much higher than those reported in other parts of the Mediterranean Sea, and elsewhere. White and transparent fragments were the most abundant form of float-ing microplastic particles. On the seafloor, microplastics in the shallow sediments were found in 90% of all sites examined, however, unlike the situation at the sea surface, the mean abundance of these particles was low $(1.21 \times 10^{-6}$ particles/m^2 $\pm 1.417 \times 10^{-6})$ in comparison to the abundances recorded in sediments in other parts of the Mediterranean. The discrepancy between the abundances of floating vs seafloor micro-plastics is not clear and may be related to the composition of the parti-cles. Polyethylene and polypropylene, the dominant marine plastics in many locations tend to float rather than sink. In addition, it is possible that sinking microplastic particles may concentrate at deeper seafloor sites and not on the fairly flat shallow sandy sediments where we sampled. The most abundant types of benthic microplastic particles were white and dark grey fragments and colored films. PCBs and PAHs were detected on plastic particles taken from the sea surface. Whereas previous work showed DDTs on microplastics in Israeli Mediterranean coastal samples, none were detected in our samples. Sixteen different types of PCBs were identified on the particles with a mean concentration of 74   ng/g; the sum of PAHs occurred at a mean concentration of 316   ng/g. The concentrations of POPs on the particles were > 3 orders of magnitude higher than in the ambient seawater. Although many of the FTIR analyses of the microplastics in this study were inconclusive ($< 75\%$ match with standard virgin plastic pellets), most of the plastic particles examined were either polyethylene or polypropylene. Whereas microplastic pollution is a cosmopolitan problem, it is especially acute in the Levant where microplastics appear to accumulate. The marine environment in the eastern Mediterranean is subject to many stressors and to the existing list of problems we need to add microplastic pollution. The challenges we face therefore are to improve our understanding of the ecological role of microplastics in the sea, and develop a strategy to address and successfully reduce this problem.

Occurrence of Microplastics in the South Eastern Black Sea

U. Aytan[1], A. Valente[2], Y. Senturk[1], R. Usta[1], B.E. Sahin[1], R. Mazlum[1] and E. Agirbas[1]
[1]Recep Tayyip Erdogan University (RTEU), Rize, Turkey [2]University of Lisbon, Lisbon, Portugal

Plastic pollution is considered one of the most urgent and difficult environmental problems in the Black Sea due to excessive river discharge of several industrialized countries into this semi-enclosed sea. However, microplastic pollution in the Black Sea region has not yet been qualified. Here, for the first time, the occurrence and distribution of microplastics are reported for the Black Sea. Microplastics were assessed from zooplankton samples taken during two cruises along the south eastern coast of the Black Sea in the November of 2014 and February of 2015. In each cruise neuston samples were collected at 12 stations using a WP2 net with 200 μm mesh. Microplastics (0.2–5 mm) were found in 92% of the samples. The primary shapes were fibers and plastic films, followed by fragments, and no micro beads were found. Average microplastic concentration in November ($1.2 \pm 1.1 \times 10^3$ par./m^3) was higher than in February ($0.6 \pm 0.55 \times 10^3$ par./m^3). Reduced concentrations in February were possibly caused by increased mixing. The highest concentrations of microplastics were observed in offshore stations during November sampling. The heterogeneous spatial distribution ($0.2 \times 10^3 - 3.3 \times 10^3$ par./m^3 for all samples) and accumulation in some stations could be associated to transport and retention mechanisms linked with wind and the dynamics of the rim current, as well by different sources of plastic. Microplastic to zooplankton ratios varied between 0.05 and 1.01 (mean 0.30 ± 0.31) in November to 0.01 and 0.58 (mean 0.20 ± 0.20) in February. These ratios indicate bioavailability of microplastics to zooplankton and many other organisms. The relatively high microplastic concentrations suggest that Black Sea is a hotspot for microplastic pollution and there is an urgency to understand their origins, transportation, and effects on marine life.

Microplastics Migrations in Sea Coastal Zone: Baltic Amber as an Example

I. Chubarenko

P.P. Shirshov Institute of Oceanology of Russian Academy of Sciences, Kaliningrad, Russia

Behavior of microplastic particles (L < 5 mm) in marine environment is complicated, and no field observations are available. Baltic amber (succinite), with its density of about 1.05–1.09 g/cm^3, fits the range of densities of slightly negatively buoyant plastics: polyamide, polystyrene, acrylic, etc. Information on amber migrations in the sea and washing of amber pieces ashore may shed some light onto general features of plastic litter/microplastic particles behavior on beaches and in sea coastal zone. Large events of "amber washing-out" at Kaliningrad sea shore typically take place in autumn-winter time, several times per year. Large pieces of amber (up to 30–40 cm long) appear on the shoreline together with patches of seaweed and marine debris after the most strong storms from northern to western directions, while smaller pieces can be found everywhere on the beaches after quite moderate winds. Amber is washed out from the depths down to 15–20 m. Massive presence in amber-containing debris of the red algae *Furcellaria lumbricalis*, dominating at depths of 6–15 m, proves this fact. Most important factors for the "amber washing-out" events are: strong and long-lasting storm, phase of wind decrease or direction change, developed long surface wind waves, and shore exposure to wind.

Characteristic wave lengths after storms, dimensions of their surf zone, and changes in underwater bottom profile were analyzed. Conclusions for plastic pieces/microplastics behavior are as follows. Slightly negatively buoyant plastic pieces should migrate for a long time between beaches and underwater slopes, being carried from the beach to the sea during storms—and moved up-slope or even washed back ashore when storms cease; this happens repeatedly until they are broken into small pieces that can be transported and deposited in deep area, out of reach of stormy waves.

Financial support of Russian Science Foundation (#15-17-10020) is acknowledged.

Pieces of Baltic amber (shown by yellow arrows) are usually washed ashore together with marine debris (wood pieces, seaweeds) and microplastic particles (the largest are highlighted by ovals).

Simultaneous Trace Analysis of Nine Common Plastics in Environmental Samples via Pyrolysis Gas Chromatography Mass Spectrometry (Py-GCMS)

Marten Fischer and Barbara Scholz-Böttcher
University of Oldenburg (ICBM), Oldenburg, Germany

There is a raising concern about plastic debris in the marine environment. The proportion of microplastic (MP) is expected to increase steadily due to ongoing fragmentation processes. Sources, distribution, and accumulation behavior of S-MP (<1 mm) is comparably poorly understood due to restricted availability of data (Cózar et al., 2014). Identification and quantification of MP in the water column, sediment, and biota is comparably time consuming and leaks of standardization. Exclusively microscopically recognition and counting are losing reliability in submicron scale. Combined microscopic and spectroscopic FTIR- and RAMAN-techniques are the most established in MP analysis. Counting their size related abundances concurrently MP particles are identified via their polymer specific spectra. Pyrolysis gas chromatography mass spectrometry (Py-GCMS) is frequently used for identification and rarely for quantification of single plastics in natural samples. Comparably fast, quantitative chemical and weight related data complementary to number and size related records are generated. Our study applies Py-GCMS combined with thermochemolysis for simultaneous analysis of nine majority plastics (PE, PP, PET, PS, PVC, PC, PA, PMMA, and PUR). Selected fragments ions of specific pyrolysis products enable a sensitive polymer specific identification and quantification on ng to μg trace level. Prior to Py-GCMS environmental samples need a multistep enzymatic, oxidative treatment and occasionally pre- or pursued density separation in order to achieve substantial MP-concentrations. The quality and potential of this method will be shown concerning reproducibility, recovery, reliability, and possible interferences with common, sometimes remaining matrix particles like wood, chitin, and cellulose fibers.

REFERENCE

Cózar, et al., 2014. Plastic debris in the open ocean. PNAS 111 (28), 10239–10244.

Extensive Review on the Presence of Microplastics and Nanoplastics in Seafood: Data Gaps and Recommendations for Future Risk Assessment for Human Health

K. Mackay, A. Afonso, A. Maggiore and M. Binaglia
European Food Safety Authority (EFSA), Parma, Italy

The volume of plastic waste in the sea is increasing, both along the coastline and in the open sea. Plastic particles have been found in fish and shellfish in various geographical locations. Substances in or attached to plastic particles like stabilizing agents, colorants, flame retardants, plasticizers, Persistent Organic Pollutant (POPs) and Polychlorinated Biphenyls (PCBs) consumed by marine organisms could migrate to the edible parts of fish and shellfish. This may increase consumer exposure to micro- and nanoplastic particles and to chemical contaminants and result in an emerging food safety issue.

The European Food Safety Authority (EFSA) was asked by the German Federal Institute for Risk Assessment (BfR) to provide a statement on the presence of microplastics and nanoplastics in food, with particular focus on seafood. This issue was discussed by the EFSA Emerging Risks Exchange Network (EREN), which is in charge of exchanging information between EFSA and the Member States on possible emerging risks for food and feed safety. In its 2013 annual report, EREN recommended EFSA to monitor the issue.

The EFSA Panel on Contaminants in the Food Chain (CONTAM) drafted the requested statement, which deals with an extensive review of the available information including previous assessments, current legislation, methods to identify and quantify microplastics and nano-plastics, sources of exposure and occurrence of microplastics and nano-plastics in food, including seafood. In addition, the main data gaps to be filled in order to perform a comprehensive assessment on the risks to human health and on the presence of microplastics and nanoplastics in food, are identified. Finally, recommendations for further research required to fill the identified data gaps are made. The statement was published in the EFSA Journal on June 23, 2016 (http://www.efsa.europa.eu/en/efsajournal/pub/4501).

State of Knowledge on Human Health Implications on Consumption of Aquatic Organisms Containing Microplastics

E.G. Gamarro and J. Toppe
Food and Agricultural Organization of the United Nations (FAO), Rome, Italy

The presence of microplastics in the marine environment has been reported repeatedly in the past years. The concern about the effects of microplastics in aquatic organisms is increasing and it is essential to take stock of the knowledge available, highlight the gaps, and indicate priority research areas to fill in the knowledge gaps in relation to human health implications of ingestion of aquatic organisms containing microplastics.

A literature review and consultation with different public institutions has been carried out to compile information about the status of microplastic levels in aquatic foods and the associated additives and contaminants. This process has revealed existing information and knowledge gaps regarding microplastic levels in marine environments, in aquatic organisms, in levels of additives and contaminants associated with microplastics, and the eventual transfer of contaminants and additives from microplastic to parts of aquatic organisms that are consumed.

In order to fill the knowledge gaps, there is a need for further research to evaluate the levels of microplastics and associated additives and contaminants in aquatic organisms containing microplastics, and the eventual effects on human health.

Effects of PVC and Nylon Microplastics on Survival and Reproduction of the Small Terrestrial Earthworm *Enchytraeus crypticus*

A. Walton[1,2], E. Lahive[2], C. Svendsen[2] and T. Galloway[1]

[1]University of Exeter, Exeter, United Kingdom [2]Centre for Ecology and Hydrology, Wallingford, United Kingdom

Microplastic litter is a pervasive environmental problem; not only in rivers and oceans but also on land surfaces and in soil. Both the presence and ecological consequences of microplastic presence to terrestrial invertebrates is poorly understood. In this study the effects of PVC and nylon particles on the survival and reproduction rate of the small (1 cm) earthworm *Enchytraeus crypticus* was observed using a series of dose−response tests over a 21-day period. Both PVC and nylon are environmentally relevant plastics; PVC used in packaging materials, medical supplies, wiring, and plumbing and is a dense particle likely to settle in soils, sediments, and sewage sludge. Nylon is commonly used as a synthetic textile, which is also likely to be found in waste water treatment sludge as fibers, with great potential for land application.

For both plastic particles, parallel concentration ranges of spiked lufa 2.2 soil were established, ranging between 12.5 g/kg and 200 g/kg. Results are presented as dose−response curves, with EC_{50} values, for both adult survival and rates of reproduction for both polymer types. A discussion of the drivers of observed results will follow with a short overview of future work required to develop the understanding of microplastics in a terrestrial context.

Microplastics: Who Is at Risk?

R. Saborowski and L. Gutow

Alfred Wegener Institute for Polar and Marine Research, Bremerhaven, Germany

Severe impacts of plastic garbage have been demonstrated for a number of marine organisms. These include, among others, entanglement, strangulation, and clogging of intestines. On the microscale, however, such devastating effects are less obvious but seem to be replaced by other adverse effects on the level of organs and cells. Ingested microplastics can induce immune responses and inflammatory cell reaction in mussels. Additionally, microplastics were shown to reduce food consumption, fecundity, and survival in copepods. However, reports about the effects of microplastics draw an inconsistent picture so far as, for example, no obvious negative effects of microplastic ingestion were observed in other species such as sandhoppers and marine isopods. Apparently, the susceptibility of an organism to the effects of microplastics depends on various factors including the habitat, feeding preferences, feeding modes, behavior, anatomy, and physiology. Suspension feeders which unselectively swallow up small particles are more vulnerable of ingesting microplastics than predators which specifically chase live prey. Moreover, anatomical features of many crustaceans, particularly the sophisticated primary and secondary filters within the stomach (figure) can prevent the passage of microscopic particles into the sensitive resorbing organs. Other taxa, such as mussels or annelids do not possess such filters and, thus, are less protected against incorporation of microplastics. Our objective is to evaluate the susceptibility of marine invertebrate species or higher taxonomic groups against microplastics with regard to their ecological, behavioral, anatomical, and physiological characteristics. This assessment is aimed to draw an overarching plot which will help to predict the risks of microplastics on the level of species and to project this knowledge to the level of taxonomic and/or functional groups and, finally, to ecosystems.

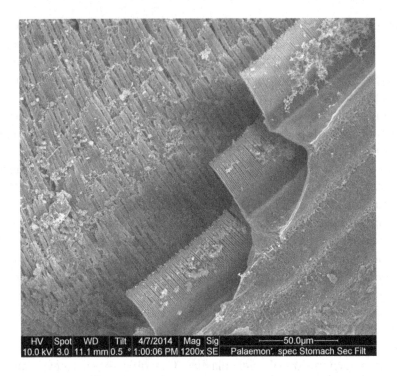

| HV | Spot | WD | Tilt | 4/7/2014 | Mag | Sig | ⊢———50.0μm———⊣ |
| 10.0 kV | 3.0 | 11.1 mm | 0.5 ° | 1:00:06 PM | 1200x | SE | Palaemon'. spec Stomach Sec Filt |

Understanding Microplastic Distribution: A Global Citizen Monitoring Effort

A. Barrows[1,2]
[1]Adventure Scientists, Bozeman, MT, United States [2]College of the Atlantic (COA), Bar Harbor, ME, United States

Understanding distribution and abundance of microplastics in the world's oceans will continue to help inform global law-making. Through recruiting and training over 500 volunteers our study has collected over 1500 samples from remote and populated areas worldwide. Samples include water collected in freshwater, at the sea surface, and throughout the water column. Surface to depth sampling has provided insight into vertical plastic distribution. The development of unique field and laboratory methodology has enabled plastics to be quantified down to 50 μm. In 2015, the study expanded to include global freshwater systems. By understanding plastic patterns, distribution, and concentration in large and small watersheds we will better understand how freshwater systems are contributing to marine microplastic pollution.

Marine Litter in the North Sea: Experiences With Monitoring

E. Leemans

The North Sea is one of the busiest sea areas when it comes to shipping, fisheries, oil & gas exploration, sand extraction, and offshore wind energy. In the summertime, tourism on beaches and on boats is popular. In all: an intensively used sea, with associated impacts such as marine pollution. Monitoring of marine litter in the North Sea area has been carried out since 25 years. From 1992−97 the marine litter on the Swedish North Sea coast was estimated at 43,000 m^3. In 2000 the OSPAR Commission started to develop a uniform monitoring methodology, which was finalized in 2010. Through this methodology and associated tools, a uniform comparison can be made on the developments of marine litter on the coastlines in the OSPAR area (http://www.ospar.org/documents?v=7260). The North Sea Foundation is one of the organizations to carry out the annual Ospar Marine Litter monitoring. This is done at exact locations on 4 different beaches. On each location, a stretch of 100 m is surveyed in detail, counting all items of litter. Additionally, on a stretch of 1 km all items >50 cm are counted. All information is uploaded in a database.

The main findings for the Dutch beaches in the period 2002−12 are: total amount of marine litter items: 60,839; top five of items: nets & ropes: 22,677; all plastic/polystyrene pieces: 11,180; all plastic bags: 3632; plastic caps/lids: 3114; and plastic crisp/sweet packets and lolly sticks: 2318.

Additional monitoring is carried out through citizen science. During a diving expedition in the North Sea, fishing gear was recovered from ship wrecks. On a total of 14 shipwrecks, 1952 kg of fishing gear was recovered, consisting of nets, fishing lines, floats, hooks, and lead.

Voluntary Beach Cleanups at Famara Beach, Lanzarote—Fighting Marine Litter Invasion and Accumulation Locally

N.A. Ruckstuhl[1]
[1]COUP, c-o-u-p.org

History: Since March 2014 citizens monthly clean the beaches around Famara on a voluntary basis. Due to the global currents Famara is a hotspot of microplastic and marine debris.

Process: This initiative was put in place by the environmental association of Famara Limpia. A variety of stakeholders are involved. The local municipality supports the beach cleanings by providing materials such as bags, gloves, and printing materials. COUP helps organizing and promoting the events, and other local associations take part in the cleanups— Participation Cuidadana, Viento del Jable, ADISLAN, Microtrofic.

Impacts: The beach of Famara is cleaner after the years of cleanup activities. Through this voluntary beach cleanups in Famara, 2–3 tons of marine debris have been collected. On an average of about 30 persons, mainly residents from aboard and tourists, participate at each cleanup, collecting roughly 70 kg of waste during each event.

Citizen science: Data about the washed up debris was collected and provided to environmental organizations and campaigns such as Ocean Conservancy, Marnoba, PlasticoCero, and Seawatchers.

Conclusion: The effectiveness of this initiative is indicated as over 3 years people were motivated to volunteer to clean the beach. However various challenges are given. Microplastic imposes a technical problem of its cleaning due to its size and to the characteristics of the location—a mix of stones, sand, and seaweeds. Without a change in our consumption behavior and production practises, cleanups need to continue. And reversely through cleanups awareness about marine debris will increase and alternatives about waste reduction can also increase. Resources, human and financial, are needed for assuring the continuity of these activities, depending heavily on the political will and its priority. An open network is essential for increasing the collaboration, transparency, and also the visibility for tackling the problem of marine debris, including microplastic.

The Wider Benefits of Cleaning Up Marine Plastic: Examining the Direct Impacts of Beach Cleans on the Volunteers

K.J. Wyles[1,2], S. Pahl[2], M. Holland[2] and R.C. Thompson[2]
[1]Plymouth Marine Laboratory, Plymouth, United Kingdom [2]Plymouth University, Plymouth, United Kingdom

Social research has demonstrated that marine environments can be psychologically beneficial to their visitors. Visits can improve individuals' well-being (how people think and feel about their lives) as well as increase awareness of marine issues; however they can also contribute to the accumulation of marine litter on the coast. Removing litter already in the marine environment is a global and challenging task. As a result, numerous clean-up initiatives exist around the world, including volunteer beach cleans where the public help to remove and/or monitor litter as part of a citizen science program (e.g., Marine Conservation Society's Beachwatch programme in the UK). Whilst these initiatives help address this issue, they may only have limited, local impact. Thus, it is important to study any broader benefits associated with these activities and whether the benefits associated with the coast remain, or are further enhanced or compromised by engaging in this proenvironmental act.

Within the allocated 5 minute session, aspects of an experimental study undertaken in the field will be reported. To examine the unique impacts of beach cleans, an experimental design was used where participants were randomly allocated to one of three structured activities: a beach clean, a rock pooling session, or a coastal walk. Participants ($n = 90$) completed surveys before and after their activity and again a week later. All activities were associated with high satisfaction and leaving with greater intentions to perform more proenvironmental behaviors; however beach cleans were rated as more meaningful and gained similarly high marine awareness as the other citizen science activity, rock pooling. Participants who engaged in the beach clean also had greater intention to volunteer in the future. Thus, this environmentally friendly act is not only beneficial for the environment, but can also be beneficial for the volunteers themselves.

"Agüita con el Plástico": Society as Part of the Solution of Plastic Pollution

J.C. Jiménez, M.S. Mederos, G. Mir and S. Rodríguez

Oficina de la Reserva de la Biosfera de Lanzarote, Lanzarote (Canary Islands), Spain

Considerable efforts have been undertaken by several social sectors such as public administrations, research teams, social and business associations, and nongovernmental organizations to mitigate the increase of single-use plastic products consumption that consequently. This fact generates problems about the increase of environmental pollution and its consequences on marine ecosystems and wild animals; in addition to the effects on public health. So, why are not those efforts showed in the consumeristic attitude of population? The global production of plastic during 2013 was around 300 million of tons and 40% of all these plastics were used to elaborate plastic packaging. (PlasticEurope, 2014). The coast of Lanzarote, Fuerteventura, and La Graciosa islands are seriously affected by plastic debris, where the most of their beaches reach a pollution concentration of microplastics greater than 100 g/L (Baztan et al., 2014). Moreover, Lanzarote's population generates around 928.64 plastic containers tons each year (Waste Department, Lanzarote's Cabildo 2014). On the other hand, the most frequent cause of mortality of the loggerhead turtle (*Caretta caretta*), one of the most common species of turtle around Canary Islands, is entanglement in fishing gear and/or plastics (Oros et al., 2016). Environmental education is considered by many authors as the best tool to manage the environment. "Agüita con el Plástico" campaign stems from the "Zero Plastic Lanzarote Biosphere Reserve" project due to the need to voice the plastic pollution problem among the population, who are an important part of the solutions. For this reason, to engage all the social sectors in finding global solutions is the principal aim of our campaign. Many actions like informative talks and info-points; presence on social networks and media; opening lines of collaboration with public and private entities; ambassadors search; etc. have been the key to success of this campaign that has crossed borders.

No Plastic Campaign Makes a Difference in Island of Principe Biosphere Reserve

M. Clüsener-Godt and A. Abreu

UNESCO, Paris, France

One year after it was first launched in February 2014, the campaign 'No plastic: a small gesture in our hands' is proving to be hugely successful. This awareness and mobilization campaign was implemented by the Island of Principe UNESCO Biosphere Reserve (São Tome and Principe) and the UNESCO Man and the Biosphere (MAB) programme to reduce plastic waste, promote access to drinking water and promote awareness on waste reduction, and management in the biosphere reserve. Fifty plastic bottles can be exchanged for a reusable, stainless steel 'Principe Biosphere Bottle' made from safe, plastic-free materials. These bottles can be replenished at various treated water points installed across the Island of Principe.

After a year-long campaign, more than 300,000 plastic bottles were removed, 13 safe water fountains were established, and 6000 'Principe Biosphere Bottles' were distributed among the local population. The campaign is the result of a partnership between the Regional Government of Principe, through the Island of Príncipe Biosphere Reserve, UNESCO's Man and Biosphere programme (MAB), the Spanish Ministry of Agriculture, Food and Environment, and the HBD group.

The Regional President of the Government of Principe announced recently, while meeting with representatives of the MAB programme and the Spanish Ministry of Agriculture, Food and Environment, that the Island of Principe would become 'plastic-free' by 2020, and that this project would serve as a model for future activities. The campaign will continue during 2016 together with a pilot project aiming to install ecopoints in several small communities.

Logistics of Coastline Plastic Cleanup and Recycling: A Literature Review and Research Opportunities

N. Brahimi[1], K. Loubar[2] and M. Tazerout[2]

[1]Rennes School of Business, Rennes, France [2]Ecole des Mines de Nantes, Nantes, France

A very recent study (Sherman and van Sebille, 2016) showed that to clean up the oceans from plastic while causing the least harm to ecosystems, the most efficient solution might be to focus on coastlines. As noted by the authors in (Sherman and van Sebille, 2016), it makes sense to collect plastic where it enters to the ocean. The issue of coastal pollution was considerably studied and several literature reviews were proposed by (Vikas and Dwarakish, 2015; Derraik, 2002; Andrady, 2011), for example.

There is more and more interest in coastline cleanup and there are many campaigns for cleanup operations (United Nations Environment Programme).

However, to the best of our knowledge, there is no study or integrated solutions on the coastline cleanup logistics, despite the obvious environmental benefits and potential economic benefits of such solutions (Hartman and Lovén, 2014). For this reason, we propose a literature review of related studies and present a framework for an integrated solution to optimize the logistics of coastline cleanup and plastic recycling by integrating data collection, data analysis, and planning decisions at the strategic, tactical, and operational levels. Recycling and economic value of collected debris are also integrated into the framework.

REFERENCES

Andrady, A.L., 2011. Microplastics in the marine environment. Mar. Pollut. Bull. 62 (8), 1596–1605.

Derraik, J.G., 2002. The pollution of the marine environment by plastic debris: a review. Mar. Pollut. Bull. 44 (9), 842–852.

Hartman, A., Lovén, E., 2014. Plastic as Marine Debris and its Potential for Economic Value. Bachelor thesis in Energy and the Environment, KTH Royal Institute of Technology, Sweden.

Sherman, P., van Sebille, E., 2016. Modeling marine surface microplastic transport to assess optimal removal locations. Environ. Res. Lett. 11 (1), 014006.

United Nations Environment Programme, Marine Litter, Beach Cleanups and Campaigns, http://www.unep.org/regionalseas/marinelitter/other/cleanups/.

Vikas, M., Dwarakish, G.S., 2015. Coastal pollution: a review. Aquat. Procedia 4, 381–388.

Tackling Microplastics on Land: Citizen Observatories of Anthropogenic Litter Dynamics Within the MSCA POSEIDOMM Project

L. Galgani and S.A. Loiselle
University of Siena, Siena, Italy

Marine litter mainly stems from terrestrial source: it is globally estimated that 80% of the debris are land based (GESAMP, 1991). In the Mediterranean Sea terrestrial debris represent 94%, of which 95% consists of plastic items, from macrofragments to microbeads (Legambiente Italy, 2015). Plastics are ubiquitous in rivers, lakes, and coastal areas and represent an emerging threat for marine and freshwater ecosystems and the regulating and provisional services they provide. However, relatively little is known about the relative abundance and sources of this anthropogenic litter. This information is fundamental to determine its impact and to identify successful mitigation strategies prior to its arrival in the marine environment.

The EU POSEIDOMM project combines citizen observatories and laboratory studies to promote the quantification and reduction of plastic pollution and explore the impact of macro and microplastics on the coastal marine environment. The project teams with FreshWater Watch, a global mass community science platform focused on developing public stewardship of aquatic ecosystems and managed by Earthwatch Institute (Oxford, UK). In 40 urban centers in Europe, Asia, North and South America, trained citizen scientists have adopted specific study sites for long term monitoring of water quality, land use, and anthropogenic litter. Data indicates a significant heterogeneity in the presence and frequency of anthropogenic litter with respect to population density, local socioeconomic characteristics, ecosystem type and season. Within the POSEIDOMM project, students and citizens adopt key rivers in Tuscany for monitoring and analysis, contributing to the online database and linking to similar community based research projects across the globe. Managing the emerging crisis of marine litter will require bottom-up participation of trained citizen scientists in accurate data acquisition and promotion of behavioral change.

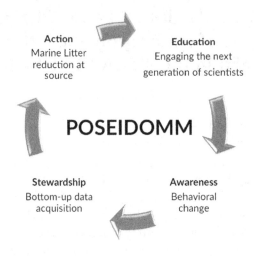

Action
Marine Litter
reduction at
source

Education
Engaging the next
generation of scientists

POSEIDOMM

Stewardship
Bottom-up data
acquisition

Awareness
Behavioral
change

REFERENCES

GESAMP, 1991. The State of the Marine Environment. Blackwell Scientific Publications, London, 146 pp.

Legambiente Italy, Plastic free sea report, http://www.legambiente.it/marinelitter/; 2015.

Environmental Science Education: Methodologies to Promote Ocean Literacy

F. Silva, L. Vasconcelos and J.C. Ferreira
New University of Lisbon, Lisbon, Portugal

Education for Environmental Citizenship is key to nature and biodiversity conservation and environmental protection. If citizens become aware of the value of natural resources then they will recognize the importance of their involvement in its protection. Therefore, educating the new and older generation citizens is the focus of our work.

Here we present the work developed under the Citizenship Component in the context of several projects, namely MARGov, MARLISCO, VoW, and Pesca Ó Peixe. These projects aim to promote active emancipatory environmental education through the discovery of the marine world, stimulating senses, imagination and, creativity.

In order to promote meaningful learning, the concept/method of problem-based learning is explored with the students to develop flexible knowledge, collaborative problem solving skills, self-directed learning, and intrinsic motivation through active learning. The multigenerational approach and the development of more appropriate, efficient, effective, and appealing learning support materials addressing the objectives of each action are found to be the key to success. The pedagogy of sharing, where everyone has something to teach and learn, reinforcing that "nobody educates nobody but we all educate each other mutually", has been systematically applied in all the projects.

The lessons learned, the methods used, and the materials developed, are now being adjusted and replicated in the Pesca Ó Peixe project. This one targets the promotion of multigenerational activities through the empowerment of young reporters for ocean literacy. The project takes benefits from the youngsters as vehicles of communication, with a strong emphasis on multigenerationality. Moreover, it empowers them through working in network with the fishing community and the media, as a way to reach out to a wider public sphere. This intends to promote awareness and knowledge of the sea, namely about marine litter and microplastics, and to create a network of continuity in the diffusion of activities and knowledge.

Microplastics, Convergence Areas, and Fin Whales in the Northwestern Mediterranean Sea

M.C. Fossi[1], C. Panti[1], T. Romeo[2], M. Baini[1], L. Marsili[1], F. Galgani[3], J.-N. Druon[4] and C. Lapucci[5]

[1]University of Siena, Siena, Italy [2]ISPRA-Institute for Environmental Protection and Research, Milazzo, Italy [3]IFREMER, Bastia, France [4]Institute for the Protection and Security of the Citizen (IPSC), Ispra, Italy [5]LaMMA Consortium-CNR Ibimet, Sesto Fiorentino, Italy

The impact that microplastics have on baleen whales particularly in semienclosed basin such as the Mediterranean Sea is a question that is as yet largely unexplored. The Mediterranean Sea is one of the area most affected by litter in the world. The highest percentage (~80%) of marine litter consists of plastics, including microplastics (particles <5 mm). Research on the impact of microplastics on large filter-feeding species, such as fin whale (*Balaenoptera physalus*), is still in its infancy. Here we present the results of the PlasticPelagos pilot project, supported by the Italian Ministry of Environment, focused to investigate the overlap between microplastic distribution and fin whale feeding ground in convergence areas (gyres) of the SPAMI Pelagos Sanctuary (northwestern Mediterranean Sea). A sampling cruise in September 2014 extending over 967 miles allowed for collecting surface microplastic samples ($n = 21$), surveying macrolitter, monitoring cetaceans, and collecting fin whales skin biopsies. Two operational models of ocean circulation and fin whale potential habitat were used to localize possible convergence areas of marine litter distribution and likely presence of foraging fin whales. A multilayer approach was used to investigate the possible overlap between microplastic convergence areas and fin whale feeding ground. The three layers of field data, microplastic abundance (items-micro/m^2), macroplastic abundance (items-macro/km^2) and cetacean presence, were compared with the two models. The plastic data set has revealed high occurrence of microplastics ($0.009-0.260$ items/m^2) in the surface neuston/plankton samples and a significant overlap with the areas showing high macroplastic density (Spearman $R = 0.6127$). Moreover, areas of high microplastic densities detected on circulation maps largely overlapped with potential fin whale feeding grounds suggesting that whales are exposed to microplastics when foraging in the Pelagos Sanctuary during summer. Ecotoxicological analysis (phthalates concentrations and biomarker responses) of the collected fin whale skin biopsies were also carried out to further support this hypothesis.

Microplastics in Marine Mesoherbivores

L. Gutow[1], A. Eckerlebe[1], J. Hämer[1], L. Giménez[2] and R. Saborowski[1]

[1]Alfred Wegener Institute Helmholtz Centre for Polar and Marine Research (AWI), Bremerhaven, Germany [2]Bangor University, Bangor, United Kingdom

Microplastics accumulate worldwide in marine habitats. Especially the surface waters of the oceans (including polar sea ice), the water column, and seafloor sediments are contaminated by huge amounts of microplastics. Accordingly, most studies on the uptake of microplastics and their effects focus on species from these specific habitats. Marine benthic mesoherbivores, which live primarily associated with seaweeds on rocky shores, have been widely ignored in studies on microplastics presumably because this functional group of organisms is believed to be at low risk of encountering synthetic particles in their natural habitats. We tested whether seaweeds can collect microplastics from the water column and make them available for ingestion by marine mesoherbivores. In laboratory experiments pieces of the brown seaweed *Fucus vesiculosus* were incubated in microplastic suspensions and subsequently screened for adhering microplastics. We used different types of fluorescent microplastics including commercial beads, fragments, and fibers. Microplastics readily adhered to the surface of the algal pieces. The particle density mostly correlated with the concentration of particles in the suspension. Tidal emergence and desiccation of contaminated algae did not enhance the adherence of the particles. In feeding assays algal pieces contaminated with microplastics were offered to the herbivorous periwinkle *Littorina littorea*. The gastropods did not prefer clean over contaminated algae indicating that they do not recognize solid non-food particles in the submillimeter size range as deleterious. In dissected periwinkles microplastics were found in the stomach and in the gut but not in the hepatopancreas, which is the principle digestive organ of gastropods. Substantial amounts of microplastics were released with fecal pellets indicating that the particles do not accumulate rapidly inside the intestines. Our results indicate that benthic seaweeds may function as a vector for the entry of microplastics into marine food webs at low trophic levels.

Investigating the Presence and Effects of Microplastics in Sea Turtles

E. Duncan[1,3], A. Broderick[1], T. Galloway[2], P. Lindeque[3] and B. Godley[1]
[1]University of Exeter, Penryn, United Kingdom [2]University of Exeter, Exeter, United Kingdom
[3]Plymouth Marine Laboratory, Plymouth, United Kingdom

The accumulation of microplastic fragments in the oceans is an emerging area of concern. Due to their small size they are available for ingestion by a wide range of organisms meaning that they could have broad scale impacts. Studies are now reporting the uptake and retention of microplastics in a range of marine organisms. This highlights the need for further research into the impacts of microplastics on marine fauna of conservation concern. While methods for investigation of macroplastic pollution in sea turtles are currently being developed, the knowledge of the effects or even the presence of microplastic ingestion within this taxon remains unknown. We tested an optimized enzymatic digestion methodology previously utilized upon zooplankton that allowed for the removal of biological material from sea turtle gut contents samples that were collected from stranded loggerhead (*Caretta caretta*; $n = 18$) and green (*Chelonia mydas*; $n = 27$) turtles from the coastline of Northern Cyprus (2011−15). This method allowed us to isolate and identify microplastics as small as 50 μm in the gut content of both species. We hypothesize on the origin of these materials and the possible impacts that they may have. Adapting a methodology, which was previously used on marine invertebrates, revealed that this novel and cryptic threat is present and possibly impacting these marine megafauna species. Additionally, it is a demonstration of a methodology that could be applied in similar investigations on other large marine vertebrates, generating essential knowledge if we are to grasp the impacts of microplastics across marine food webs.

Microplastics Presence in Sea Turtles

P. Ostiategui-Francia[1], A. Usategui-Martín[1] and A. Liria-Loza[2]
[1]University of Las Palmas de Gran Canaria (ULPGC), Las Palmas de Gran Canaria, Spain
[2]Asociación para el desarrollo sostenible y biodiversidad (ADS Biodiversidad)

Marine pollution is a major threat for marine ecosystems and plastic pollution is being heavily studied and widely reported. Currently, microplastic contamination has moved into the focus of environmental research, with numerous studies addressing the occurrence of microplastics in the water column, in sediments, and in filter-feeding organisms. A major challenge is to identify the effects of microplastics on wild fauna, as well as to understand the trophic transfer of the particles among trophic levels. Several studies have reported on the presence of microplastics in marine vertebrates such as fulmars and fin whales, implying a direct uptake, or trophic transfer to the highest levels of the food web. In this study, the presence of microplastics in marine turtles was analyzed for first time in both, wild turtles—stranded and sent to the Gran Canaria Wildlife Recovery Center—and captive turtles—from the loggerhead reintroduction program and reared in captivity. Within this work we developed a new methodology that allows for identifying microplastic in live animals without disturbing them, through feces analysis. Microplastics were extracted from turtle feces by immersion in hypersaline water, filtered through a 0.7 μm paper filter MN − GF − 1 disposed in a Swinnex-25 syringe vacuum filter and classified according to Marine Litter Guidance categories. Three types of microplastics were found in feces samples: filaments, spheruloid pellets, and plastic fragments. No significant differences in microplastics presence were observed between wild and captive turtles, although macroplastics were found only in wild animals. However, microplastic densities varied with food type and origin. This study is the first record of microplastics in loggerhead turtles. The results suggest that the presence of microplastics in turtles is directly related with food ingestion indicating transfer of microplastics along the food chain, rather than direct uptake or degradation of macroplastics ingested.

REFERENCES

Baztan, J., Carrasco, A., Chouinard, O., Cleaud, M., Gabaldon, J.E., Huck, T., et al., 2014. Protected areas in the Atlantic facing the hazards of micro-plastic pollution: first diagnosis of three islands in the Canary Current. Mar. Pollut. Bull. 80, 302−311.

Directive, S.F., 2013. Guidance on Monitoring of Marine Litter in European Seas.

Fossi, M.C., Panti, C., Guerranti, C., Coppola, D., Giannetti, M., Marsili, L., et al., 2012. Are baleen whales exposed to the threat of microplastics? A case study of the Mediterranean fin whale (Balaenoptera physalus). Mar. Pollut. Bull. 64, 2374−2379.

Galgani, F., Claro, F., Depledge, M., Fossi, C., 2014. Monitoring the impact of litter in large vertebrates in the Mediterranean Sea within the European Marine Strategy Framework Directive (MSFD): Constraints, specificities and recommendations. Mar. Environ. Res. 100, 3−9.

Ivar do Sul, J.A., Costa, M.F., 2014. The present and future of microplastic pollution in the marine environment. Environ. Pollut. 185, 352−364.

Lusher, A.L., McHugh, M., Thompson, R.C., 2014. Occurrence of microplastics in the gastrointestinal tract of pelagic and demersal fish from the English Channel. Mar. Pollut. Bull. 94−99.

Ryan, P.G., Moore, C.J., Franeker, J.A. van, Moloney, C.L., 2009. Monitoring the abundance of plastic debris in the marine environment. Philos. Trans. R. Soc. Lond. B Biol. Sci. 364, 1999−2012.

Sanchez, W., Bender, C., Porcher, J.-M., 2014. Wild gudgeons (Gobio gobio) from French rivers are contaminated by microplastics: preliminary study and first evidence. Environ. Res. 128, 98−100.

Thompson, R.C., Olsen, Y., Mitchell, R.P., Davis, A., Rowland, S.J., John, A.W.G., McGonigle, D., Russell, A.E., 2004. Lost at sea: where is all the plastic? Science 304, 838.

Valente, A.L., Marco, I., Parga, M.L., Lavin, S., Alegre, F., Cuenca, R., 2008. Ingesta passage and gastric emptying times in loggerhead sea turtles (Caretta caretta). Res. Vet. Sci. 84, 132−139.

Factors Determining the Composition of Plastics From the South Pacific Ocean—Are Seabirds Playing a Selective Role?

V. Hidalgo-Ruz[1,2], H. Frick[3], M. Eriksen[4], D. Miranda-Urbina[1,2], C.J.R. Robertson[5], R.P. Scofield[6], C.G. Suazo[7], G. Luna-Jorquera[1,2,8], M.M. Rivadeneira[8] and M. Thiel[1,2,8]

[1]Universidad Católica del Norte, Coquimbo, Chile [2]Millennium Nucleus of Ecology and Sustainable Management of Oceanic Island (ESMOI), Coquimbo, Chile [3]University of Copenhagen, Frederiksberg, Denmark [4]Five Gyres Institute, Los Angeles, CA, United States [5]Wellington, New Zealand [6]Canterbury Museum, Christchurch, New Zealand [7]Justus Liebig University Giessen, Giessen, Germany [8]Centro de Estudios Avanzados en Zonas Áridas (CEAZA), Coquimbo, Chile

Plastic items can be found throughout the ocean, where they interact with a number of external agents, for instance plastic uptake by seabirds. Since seabirds use visual mechanisms for feeding, plastics with certain characteristics might have a higher likelihood of uptake. In the present study we focus on several factors (density, radiation, uptake by seabirds) to understand how these affect the composition of plastic marine debris. We characterized plastic samples from the following compartments: (1) Continental Chilean beaches ($n = 1456$ plastic items) (2) Easter Island beaches ($n = 1008$), (3) South Pacific Gyre ($n = 1120$), (4) surface-feeding seabirds ($n = 3511$) (5) diving seabirds ($n = 722$), (6) seabird nesting material ($n = 1640$); according to their type, specific density, color, and shape. Hard plastic items were the most abundant type, except for Chilean beaches (mostly expanded polystyrene). Specific density of most plastic items was lower than seawater (>1.025 g/mL), although 11% of plastics from Chilean beaches were non-buoyant. Plastics in this compartment had the highest diversity of colors (mainly blue, white/gray, yellow, and red). White/gray was the dominant color in all other compartments, except for nests which contained mostly red items. Plastics from continental beaches were most distinct from all others, which was mainly explained by plastic types (EPS and hard plastics) and colors (white/gray and blue). Plastics from the Gyre and Easter Island beaches were relatively similar, while seabird samples formed a separate, yet divergent group (plastics ingested by surface-feeders and divers being different from those found in nesting areas). These results suggest that seabirds might take up certain plastics, which seems to be especially the case for plastics found in nests and nesting areas. This highlights the important role of biological processes in shaping the composition of marine plastics and the impacts that this can have on seabirds and other organisms frequently confronted with oceanic plastics.

Microplastics and Marine Mammals: Studies From Ireland

A. Lusher[1,2], G. Hernandez-Milian[3], S. Berrow[4], E. Rogan[3] and I. O'Connor[2]

[1]National University of Ireland Galway, Galway, Ireland [2]Galway-Mayo Institute of Technology, Galway, Ireland [3]University College Cork, Cork, Ireland [4]Irish Whale and Dolphin Group, Kilrush, Ireland

Globally, microplastics are an emerging marine pollutant. Once in the environment, microplastics become available for interaction with a number of marine organisms. It is extremely hard to observe the interactions of larger marine organisms with microplastics. In this study we present findings from ongoing collaborative research from Ireland which utilizes the Irish Whale and Dolphin Group and the University College Cork (UCC) stranding schemes. This project developed methods for the identification of microplastics, and confirmed microplastics identity using FT-IR. By following dissection protocols which have been optimized to reduce contamination, this study identified microplastics presence in all marine mammal species that were analyzed. The method utilizes dietary analysis for items >1 mm, and 10% KOH to dissolve remaining organic material <1 mm−150 μm. Also a review of incidences of plastic (macro and micro) reported in the digestive tracts of stranded and by-caught marine mammals on the Irish coasts from 1990−2015 was carried out. Using the dissolving method, all animals from seven different species contained microplastics. The review found that 11 of the 21 species ($n = 38$) that had stranded on the Irish coast (1990−2015) contained macro- and/or microplastics. The results of this study have been combined with data from diet to infer the potential for plastics to transfer between predators and prey. This is the first report of microplastics in a diverse group of apex predators from the marine environment. Further studies need to address the impacts that these plastics could have on organisms if they are retained.

Primary (Ingestion) and Secondary (Inhalation) Uptake of Microplastic in the Crab *Carcinus maenas*, and Its Biological Effects

A. Watts[1], M. Urbina[2], C. Lewis[1] and T. Galloway[1]

[1]University of Exeter, Exeter, United Kingdom [2]Universidad de Concepción, Concepción, Chile

Ingestion of microplastics has been demonstrated in a range of marine animals both in the laboratory and in the oceans, but we are only just starting to understand how microplastics may be transferred across trophic levels or different methods of uptake. Here we use two simple models (ingestion and inhalation model) to investigate the bioaccumulation, retention, and biological effects of microplastics.

We used an ingestion model comprised of a trophic link between mussels (*Mytilus edulis*) and crabs (*Carcinus maenas*). Our inhalation model used crabs exposed to microplastics via the water, allowing them to be taken up via inhalation through the gill chamber. We used both microspheres of 8−10 microns in diameter and microfibers of 1−5 mm in length.

Following ingestion of contaminated mussels, microplastics were rapidly distributed across body tissues, and retained for over 20 days in the foregut and gills of the crabs. Ingestion of polypropylene rope microfibers (1−5 mm in length) was associated with a significant reduction in the crabs' energy budget, and mechanical degradation of the fibers. Inhalation of microspheres had transient but limited effects on oxygen consumption and osmoregulatory capacity of the gills.

These results illustrate that crabs can both ingest and inhale microplastics. Ingestion of microfibers has a significant effect on fitness and the energy budget, and can alter the physical nature of the microplastics themselves. Gill accumulation is important for further trophic transfer, but no negative impacts were seen in the crab under the conditions of these assays. These results add important information on the biological effects of microplastics in an important coastal omnivore.

Plastic in Atlantic Cod (*Gadus morhua*) From the Norwegian Coast

D.P. Eidsvoll[1], I.L. Nerland[1], C.C. Steindal[2] and K.V. Thomas[1]

[1]Norwegian Institute for Water Research (NIVA), Oslo, Norway. [2]University of Oslo, Oslo, Norway

Microplastics contaminate oceans and affect marine organisms in several ways. This study documents the occurrence of microplastic (<5 mm), mesoplastics (5–20 mm) and macroplastic (20+ mm) in the stomachs of cod (*Gadus morhua*), one of the most common and economically important marine fish in Norway. Fish stomachs ($n = 302$) were examined from six different locations along the coast of Norway. Ten individual stomachs contained items in them identified as synthetic polymers, eight of these were from a single location (Bergen). All objects found in the stomachs analyzed by Fourier Transform Infrared Spectroscopy (FTIR) scanned and subsequently compared with FTIR libraries to confirm the identity of the items.

Preventing contamination, all work with the stomachs was performed under strict rules in a special clean pathology laboratory. Ingested items, not resembling natural stomach content were removed, scanned using FTIR and subsequently compared with FTIR libraries.

This study provides a record of plastic polymers being identified in the stomachs of cod in three out of six locations. We found polypropylene, styrene and polystyrene, polyvinyl chloride, PMMA or Plexiglas, polyethylene terephthalate, polytetrafluoreten (Teflon), and nylon 66. Common fibers seen in other studies (1–2 mm) were not present in our samples. Plastic items were more prevalent inside fish with a full stomach content versus those with empty stomachs.

Extraction and Characterization of Microplastics in Marine Organisms Sampled at Giglio Island After the Removal of the Costa Concordia Wreck

C.G. Avio[1], S. Gorbi[1], L. Cardelli[1], D. Pellegrini[2] and F. Regoli[1]

[1]Università Politecnica delle Marche, Ancona, Italy [2]Istituto Superiore per la Protezione e la Ricerca Ambientale, ISPRA, Livorno, Italy

In this study the presence and characterization of microplastics (MPs) was assessed in several species collected at the Giglio Island during the parbukling project of the Costa Concordia wreck. This scenario was chosen because affected by the presence of huge engineering activities, thus providing an interesting area to study the possibility to use marine organisms as indicators of MPs pollution.

Benthic fish were collected in proximity of the stern of the wreck and in a control site during the summer 2014, while mussels were translocated at two depth (-5 and -40 m) and in three different seasons (winter and spring 2013, summer 2014) at different distance from the wreck. A recently validated extraction protocol was applied to extract MPs from gastrointestinal tracts of fish and soft tissues of mussels. Extracted particles were characterized in terms of size, shape, and polymer typology trough microscopy and FT-IR analyses.

Results showed that fish were highly susceptible to MPs ingestion: extracted MPs were mostly represented by fragments and lines, suggesting a possible relationship with human activities related to the wreck removal. Transplanted invertebrates typically exhibited a lower frequency of MPs in soft tissue and the translocation period did not allow to highlight significant differences between sites at different distance from the wreck. On the other hand, a higher ingestion of MPs was observed in surface- compared to bottom-transplanted specimens, and in summer compared to winter and spring experiments, suggesting a different distribution and presence of MPs in the marine environment.

In conclusion, this study provides new insights on the presence and typology of MPs in marine organisms, representing an example of biomonitoring to evaluate MPs anthropogenic pollution in the marine environment.

Floating Plastic Marine Debris in the Balearic Islands: Ibiza Case Study

M. Compa[1], J.M. Aguiló[2] and S. Deudero[1]

[1]Instituto Español de Oceanografía, Palma de Mallorca, Spain [2]Conselleria de Medi Ambient, Agència Balear de l'Aigua i la Qualitat Ambiental (ABAQUA), Palma de Mallorca, Spain

Marine debris is ubiquitous throughout all marine and coastal ecosystems, vastly impacting both ecological and biological conditions creating a need for global action. The present work constitutes a preliminary assessment of the spatial and temporal distribution of floating plastic marine debris along the coastline of the Balearic Island of Ibiza during the varying months of June–September from 2005 to 2015 collected during the annual marine debris removal and monitoring program coordinated by ABAQUA (Agència Balear de l'Aigua I Qualitat Ambiental) in the Balearic Islands.

A total of 34 months were included this long-term monitoring survey assessment. The average amount of marine debris collected each month ranged from 0.98 ± 5.01 kg/ha (August 2007) to 5.57 ± 14.33 kg/ha (August 2015). Throughout the survey, the most amount of overall plastic collected during the summer period was during 2015. These initial results from the ongoing marine debris removal and monitoring program indicate marine plastic is a persistent pollutant in the marine environment and increased research is necessary to make informed decisions in policy regarding plastic marine pollution.

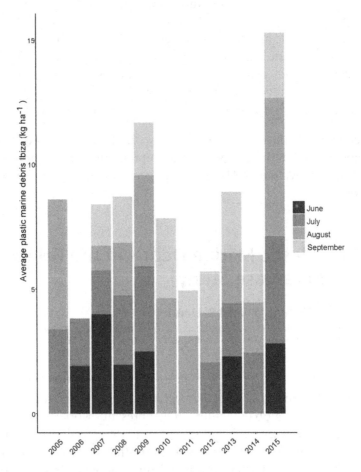

Summary of average annual marine debris (kg/ha) collected along the island of Ibiza from 2005−15.

Enzymes—Essential Catalysts in Biodegradation of Plastics

V. Perz[1], K. Haernvall[1], D. Ribitsch[1,2] and G.M. Guebitz[1,2]
[1]Austrian Centre of Industrial Biotechnology ACIB, Tulln, Austria [2]BOKU University of Natural Resources and Life Sciences, Vienna, Tulln, Austria

For further information on this study, please contact the authors.

Ireland's Microplastics: Distribution in the Environment and Interactions With Marine Organisms

A. Lusher[1,2]
[1]National University of Ireland Galway, Galway, Ireland [2]Galway-Mayo Institute of Technology, Galway, Ireland

Whilst research has established that microplastic is an emerging contaminant around the globe, its effects remain largely unknown and its associated risks unquantified. Addressing these issues is urgent as microplastic pollution will become more abundant through increased use and continued disposal of larger plastics items. During a three year project, microplastic distribution and interaction with marine biota was studied. Firstly, a novel technique was developed for continuous monitoring of microplastics in subsurface waters. Using this method, microplastics were found throughout the northeast Atlantic in 94% of the samples (average 2.46 particles/m^3). Results suggest that their distribution may be related to environmental conditions, including wind speed and sea-surface temperature. Secondly, the association of microplastics and oceanographic features was investigated during transatlantic crossings. It is hypothesized that the rotation patterns of mesoscale eddies could cause the propagation of suspended microplastics across the North Atlantic with a potential for accumulation in higher latitudes. The North Atlantic supports a large variety of economically and ecologically important marine organisms, including fish and megafauna. In this study the ingestion of microplastics by mesopelagic fish was investigated. Approximately 11% of the 761 fish examined had microplastics present in their digestive tracts. No clear difference in ingestion frequency was identified between species, location, migration behavior, or time of capture. While ingesting microplastic may not negatively impact individual mesopelagic fish, the movement of mesopelagic fish from the euphotic zone to deeper waters could mediate transfer of microplastics to otherwise unexposed species and regions of the world's oceans. This initial assessment of the distribution and interactions of microplastics provides a foundation for assessing the ecological risks posed by microplastics in the North Atlantic. Further work, including modeling, to assess the likelihood and expected impact of microplastic interactions with marine biota is required.

Deposition of Microplastics in Marine Sediments From the Irish Continental Shelf

J. Martin[1], A.L. Lusher[1], R.C. Thompson[2] and A. Morley[1]

[1]National University of Ireland Galway, Galway, Ireland [2]Plymouth University, Plymouth, United Kingdom

It is assumed that plastics have a long residence time in marine environments due to their durability and recorded accumulation within oceans and benthic sediment. Here we present high-resolution analysis of box cores to produce a timeline of microplastic deposition on the Irish continental shelf. This is the first documentation of microplastic deposition in Irish marine sediments. Sediment cores were obtained from cruises on board of the RV Celtic Explorer and RV Celtic Voyager between 2014 and 2015. They cover the important commercial benthic fishery of the Aran Grounds, Galway, and remote sites off the coast of Mayo. Microplastics were recovered and classified using visual and chemical techniques. Fourier Transform Infrared Spectroscopy allowed for identification of polymer types and the control of potential contaminants. Microplastics were most abundant (66%) within the water−surface interface of the seafloor and within surface sediments (top 1 cm). In the Aran Grounds microplastics were found to a depth of 2.5 ± 0.5 cm within the sediments studied. Assessing cores from the Aran Grounds with radio carbon dating (AMS 14C) provided sedimentation rates for the area of 53 years per 0.5 cm and showed disturbances, including bioturbation and fishing practices, may influence the distribution of microplastics within the sediment column.

Where Go the Plastics? And Whence Do They Come? From Diagnosis to Participatory Community-Based Observatory Network

J. Baztan[1,2], E. Broglio[2,3], A. Carrasco[4], O. Chouinard[2,5], F. Galgani[6], J. Garrabou[2,3], T. Huck[2,7], A. Huvet[8], B. Jorgensen[2,9], A. Liria[2,10,13], A. Miguelez[4], S. Pahl[11], I. Paul-Pont[12], R. Thompson[11], P. Soudant[12], C. Surette[2,5] and J.-P. Vanderlinden[1,2]

[1]Université de Versailles SQY, Guyancourt, France [2]Marine Sciences For Society [3]Institut de Ciències del Mar, Barcelona, Spain [4]Observatorio Reserva de Biosfera, Arrecife, Spain [5]Université de Moncton, Moncton, NB, Canada [6]IFREMER, Bastia, France [7]UBO-CNRS-LPO, Brest, France [8]IFREMER, Plouzané, France [9]Cornell University, Ithaca, NY, United States [10]Asociación para el desarrollo sostenible y biodiversidad (ADS Biodiversidad) [11]Plymouth University, Plymouth, United Kingdom [12]IUEM, CNRS/UBO, Plouzané, France [13]University of Las Palmas de Gran Canaria (ULPGC), Las Palmas de Gran Canaria, Spain

Plastics: wonder products facilitating our daily lives thanks to their usefulness. From the point of view of producers, plastics are the material for the twenty-first century; from the point of view of sea turtles, plastics cause 30% of their deaths. Plastics are a crossroads where the complexity of multiple rationality, power, and ethics intersect and can be better understood.

Since our first microplastics sampling diagnosis campaign in 2008 to the implementation of the participatory community-based observatory network, we have developed two main work processes that feed further actions along protected shores contaminated with plastic in the Atlantic and Mediterranean systems. Working from the perspective of the EU Marine Strategy Framework Directive and complementary legislative frameworks, the two main processes are:

- Diagnosis: From the majority of the 1000 studied samples, we established a microplastic pollution vulnerability baseline of studied beaches, we improved the sampling methodology, and we identified the need for long-term data series.
- Long-term observation: Through the participatory community-based observatory network pilot sites of Lanzarote, Barcelona, Lugo, Finistère and Maine, we started long-term data collection that allows us to identify and better understand the seasonal variability of microplastic pollution and establish pollution-level references to track changes over time and space. The validation of the data from these five pilot sites allows us to move to the next step of working with local stakeholders to construct a more robust network of sites along all identified vulnerable shores.

Our results show:

- The key role of local stakeholders in the participatory observatory network, connecting science-based processes with community values.
- The strong temporal variability confirmed from the last three years of time series data collected at the Finistère sampling station and validated by preliminary data analysis from Lanzarote, Lugo, Barcelona, and other stations in the network.
- The high concentrations of PCBs found in Lanzarote's Famara Beach samples confirm the hazardous aspect of microplastics; they carry toxic chemicals to remote and pristine ecosystems. Famara's samples reach in-pellet PCB concentrations of 7.4 ng/g-pellet for CB-138, with a total cumulative PCB value of 31.5 ng/g-pellet.

How can we truly solve the problem?

Distribution and Composition of Microplastics in Scotland's Seas

M. Russell and L. Webster
Marine Scotland – Science, Scottish Government, Aberdeen, Scotland

Microplastics are ubiquitous in the marine environment. They can be ingested by many marine organisms, causing adverse effects in the organism. In addition microplastics can potentially adsorb and concentrate pollutants which can be transferred to the organism, with the potential to biomagnify up the food chain. Surface water sampling has taken place in Scotland's seas since 2011 to assess whether microplastics posed a problem. Sea surface waters were sampled initially using a bongo net but since 2014 a 333 µm neuston net mounted on a catamaran swimmer body (Neuston Net acc. to David/Hempel Model 300, supplied by Duncan & Associates, Cumbria, England) has been employed. This can be towed at ship speeds of up to 8 knots, increasing the area covered on the research cruises.

A preliminary separation of collected microplastics showed that almost 72% of microplastics collected over 6 years were particles (varying shapes and colors) and 28% were fibers from fishing nets and lines. Fourier Transform Infrared Spectroscopy (FTIR) was used to confirm identification of suspected microplastics by comparing the spectra of the microplastic to those in a commercial library. For the years 2011–15 the count of microplastics confirmed by FTIR remains relatively constant over the period. The numbers of particles and fibers collected in January 2016 was almost three times that collected in total in the previous 5 years. This could be related to the storms experienced in the weeks leading up to the cruise, with more microplastics being resuspended from the shoreline in the high energy storm conditions. Of the areas sampled in 2016 the Sea of Hebrides and the Solway had the highest numbers of microplastics. One significant problem with FTIR is that the libraries are generally of pristine plastics whereas microplastics from the sea are generally weathered. Work is currently underway at Marine Scotland Science to obtain spectra for artificially weathered plastics and to build a library of these.

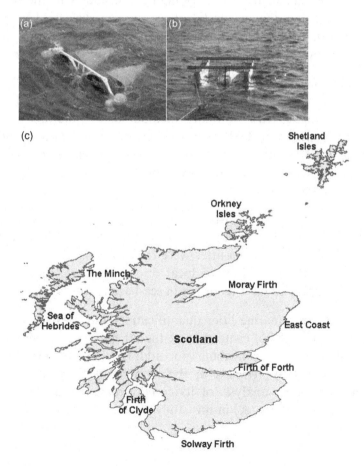

(a) Bongo net, (b) neuston net with catamaran swimmer body and (c) areas sampled in 2016.

Marine Litter Accumulation in the Azorean Archipelago: Azorlit Preliminary Data

J.P.G.L. Frias[1], R. Carriço[1], Y. Rodriguez[2], N. Rios[2], S. Garcia[3] and C. Pham[1]

[1]Universidade dos Açores, Horta, Portugal [2]OMA – Observatório do Mar dos Açores, Horta, Portugal [3]DRAM-Direção Regional dos Assuntos do Mar, Horta, Portugal

Worldwide awareness concerning environmental impacts associated with marine litter, particularly microplastics, have risen in recent decades, breaking way to new scientific research approaches and policy making decisions to address and minimize the problem caused by these materials.

The lightweight of marine litter highly contributes to the distribution and accumulation in coastal areas and sea surface. The Azores archipelago (north-eastern Atlantic) is particularly prone to marine litter accumulation due to its proximity to the North Atlantic Gyre.

In order to evaluate litter accumulation on coastal areas, 42 beaches across the archipelago were sampled, between February and March of 2016, according to two sampling methodologies (microplastics and Oslo-Paris Convention (OSPAR)), with the goal to identify accumulation zones, types, and densities of macro and microlitter. The campaign results and data analysis were presented for the first time at MICRO 2016 conference.

Different litter types (plastic, glass, metal, paper, and others) and high density variabilities both on macro (0.008 to 19.5 items/m^2) and micro (0 to 666.5 items/m^2) litter were found for the Azores archipelago beaches. Although litter of local origin was occasionally found, most items appear to have its origin in sea-based sources.

This is the first marine litter quantification study that covers all of the Azores archipelago, using state-of-the-art beach sampling methodologies that highly contribute to address the Marine Strategy Framework Directive (trends in the amount of litter deposited on coastlines, including analysis of its composition, spatial distribution, and where possible, source) in this study region.

Microplastics in the Adriatic—Results from the DeFishGear Project

A. Palatinus[1], M.K. Viršek[1] and A. Kržan[2]

[1]Institute for Water of the Republic of Slovenia, Ljubljana, Slovenia [2]National Institute of Chemistry, Ljubljana, Slovenia

The DeFishGear (IPA Adriatic, www.defishgear.org) strategic project is the first comprehensive, subregional project to deal with marine litter (macrolitter, microplastics (MP) and derelict fishing gear) in the Adriatic Sea. In the three years of the project we were able to establish a regional network of experts and institutions with knowledge and capacity to perform MP monitoring and characterization.

Using a uniform methodology, elaborated within the project, repeated sampling was performed at 34 sea-surface and 9 beach locations to give the first overall evaluation of MP pollution in different compartments/type of samples (sea surface, beach sediments, and biota).

Results from the sea surface samples indicate relatively high concentrations (average 254×10^3 particles/km^2), with the highest result in the northern Adriatic (795×10^3 particles/km^2). Beach sediments were analyzed for small MP (0.3−1 mm: average 1133 particles/kg sediment) and large MP (1−5 mm: average 113 particles/km^2). A limited analysis of MP in biota showed presence of low numbers (below 3 particles/individual). Results of analysis in all compartments are organized in a GIS database (defishgear.izvrs.si/defishgearpublic/).

Locations of sea-surface microplastics sampling.

To complement the field analyses and to identify potential concentration points a model was developed for the movement of marine litter (Liubartseva et al., 2016). Results point to strong seasonal effects that were found to be in agreement with measured MP concentrations.

A final assessment of the results will be used to develop recommendations for action.

REFERENCE

Liubartseva, S., Coppini, G., Lecci, R., Creti, S., 2016. Regional approach to modeling the transport of floating plastic debris in the Adriatic Sea. Mar. Pollut. Bull. 103, 115–127.

Floating Microplastics in the South Adriatic Sea

G. Suaria[1], C.G. Avio[2], G. Lattin[3], F. Regoli[2] and S. Aliani[1]

[1]CNR-ISMAR, La Spezia, Italy [2]Università Politecnica delle Marche, Ancona, Italy [3]Algalita Marine Research and Education, Long Beach, CA, United States

Neustonic microplastic abundance and polymeric composition were determined after a cruise conducted in the Southern Adriatic Sea between May 9 and 17, 2013. Plankton samples were collected using a neuston net (200 μm mesh size) towed at ~ 2 kts for 5−6 minutes. In the laboratory, plastic particles were hand-picked using a dissecting stereomicroscope, counted, weighed, and assigned to seven different size classes. Because of the high risk of contamination synthetic fibers were removed from our dataset and not considered in density calculations. On a subset of collected particles >700 μm ($n = 869$), FT-IR analyses were performed to characterize their polymeric

Plastic was found in all samples ($n = 30$). A total of 5163 particles were collected during the survey, the vast majority of which were irregular hard plastic fragments (97.4%).

Plastic abundance markedly increased with decreasing size indicating very high fragmentation rates (i.e., 29.4% of all particles were <300 μm and 53.8% were <500 μm). Overall, an average concentration of 0.83 ± 1.05 particles/m^2 and 485.07 ± 1153.07 g/km^2 was observed throughout the study area, with plastic abundances ranging from 0.04 particles/m^2 to a maximum of 4.65 particles/m^2.

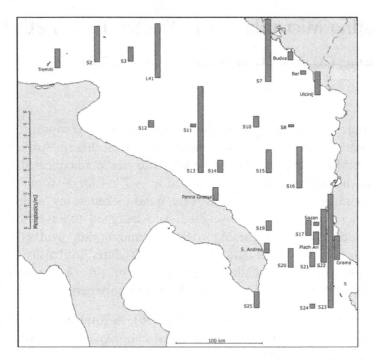

Map of the study area showing the location of all sampling stations and measured microplastic concentrations expressed as number of items/m².

Sixteen different polymer classes were identified through FT-IR analyses.

Polyethylene was the predominant polymer (41.2%), followed by synthetic paints (10.6%), polypropylene (9.7%), polyamides (8.4%), PVC (8.1%), polystyrene (4.8%) and polyvinyl-alcohol (2%). Other polymers encountered less frequently included PET, cellulose acetate, polyisoprene, poly(vinyl stearate), ethylene-vinyl acetate, epoxy resin, paraffin wax, and 11 fragments of polycaprolactone, a biodegradable polyester.

On the whole, very high levels of plastic pollution were found in the study area. Despite any clear geographical pattern was identified, the conspicuous heterogeneity in plastic distribution and polymeric composition seem to confirm the existence of multiple pollution sources insisting on the Adriatic Sea.

Implementation of the Spanish Monitoring Program of Microplastics on Beaches Within the Marine Strategy Framework Directive. First Phase

J. Buceta[1], M. Martinez-Gil[2], R. Obispo[1,2] and M. Plaza[1,2]

[1]Centro de Estudios de Puertos y Costas. CEDEX, Madrid, Spain [2]Division para la Proteccion del Mar. Directorate General for the Sustainability of the Coast and the Sea, Pza San Juan de la Cruz s/n, Madrid, Spain

The Spanish Monitoring Program for the implementation of the Marine Strategy Framework Directive (MSFD), reported to the EC on March 2015, included the subprogram BM-6 (plastic microparticles on beaches), which will collect information on the number and mass of microplastics counted on sand samples from a representative number of beaches in each of the five Spanish marine subdivisions.

As a basis for the selection of monitoring stations, the already ongoing subprogram BM-1, Marine Litter on beaches, was considered. It had been developed from 2000 in some beaches within the OSPAR Pilot Project on Beach Marine litter and was extended in 2013 to cover not only the North-East Atlantic but also the Mediterranean Sea and the Canary Islands. At present it includes26 beaches, as shown in the figure. Following the OSPAR sampling Protocol, each beach is sampled four times per year in two fixed size sections (100 and 1000 m).

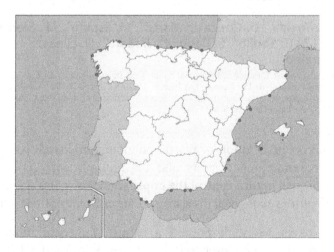

The approach followed in BM-6 development was the possible correlation between the macrolitter presence and the microliter occurrence. Taking into account the high number of beaches included in the macrolitter subprogram, it was necessary to conduct a first phase of research (main subject of the present paper) in order to reduce it and select the most adequate beaches for the routine sampling of microplastics.

Considering the sediment characteristics, some of the initial beaches were discarded attending to a clear indication on the physical impossibility for microparticles deposit (coarse sand or gravel). Thus, the number of beaches was reduced to 21. Furthermore, during the implementation of the work and taking into account the knowledge of massive occurrence of microplastics in other beaches not previously considered, additional beaches were sampled and analyzed. The methodology used, both for sampling and analysis, was based in the recommendations by the MSFD Technical Subgroup on Marine Litter for intertidal beach sediments. However, considering some limitations detected, it was necessary to slightly modify the recommended protocol. The presentation summarizes the improved protocol finally used and the main results obtained.

Operational Forecasting as a Tool for Managing Pollutant Dispersion and Recovery

A. Sánchez-Arcilla[1], D. González-Marco[1], M. Espino[1], J.P. Sierra[1], E. Álvarez[2], M.G. Sotillo[2], S. Capella[3], J. Mora[4], O. Llinás[5] and P. Cerralbo[1]

[1]Universitat Politècnica de Catalunya, Barcelona, Spain [2]Ente Público Puertos del Estado, Madrid, Spain [3]Autoridad Portuaria de Las Palmas, Las Palmas de Gran Canaria, Spain [4]Autoridad Portuaria de Santa Cruz de Tenerife, Santa Cruz de Tenerife, Spain [5]Plataforma Oceanica de Canarias (PLOCAN), Telde – Gran Canaria, Spain

The fate and concentration of plastic pollutants in coastal seas depends on meteo-oceanographic factors and the multiple entry points in the domain. This will in turn control the implications for water quality, transmission of various biological diseases and the efficiency of recovery.

In this paper we shall present a robust numerical model linked sequence that has been prepared jointly by Puertos del Estado and

LIM-UPC. It is based on advanced, coupled meteo-oceanographic models that account for the main physical mechanisms responsible for circulation and dispersion. From here concentrations, times of renewal, and trajectories can be derived with enough resolution so as to capture the important gradients in topography and bathymetry characteristic of coastal seas. The validation of these forecasts will require suitable data that address the various fields numerically predicted. This includes wind, atmospheric pressure, waves, currents, and also the variations in mean water level. Such an observational effort needs also initial and boundary conditions that for dynamically narrow shelfs can shape the water, sediment, and pollutant fluxes in the platform. In the paper we shall discuss the nested sequence that has been prepared and is now run preoperationally for a set of Canary Island harbors, where the Plocan data can provide valuable boundary information for model assessments.

The simulated current dispersion patterns will be the basis to discuss impacts and recovery policy so as to maintain a water quality status in harbor areas and adjacent coastal tracts that comply with regulations and the requirements of coastal users.

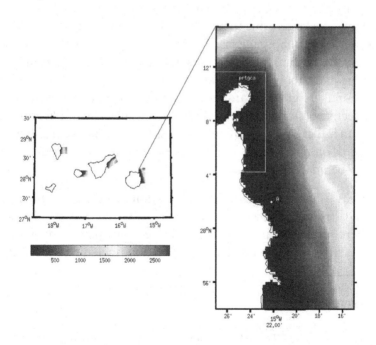

Diagram of the Canary Island harbors considered in the simulations. Showing also the nested grids and the Plocan station (A).

What Do We Know About the Ecological Impacts of Microplastic Debris?

C.M. Rochman[1], M.A. Browne[2,3], A.J. Underwood[4], J.A. van Franeker[5], R.C. Thompson[6] and L.A. Amaral-Zettler[7,8]

[1]University of California, Davis, CA, United States [2]University of California, Santa Barbara, CA United States [3]University of New South Wales, Sydney, NSW, Australia [4]University of Sydney, Sydney, NSW, Australia [5]Institute for Marine Research and Ecosystem Studies IMARES, Texel, The Netherlands [6]Plymouth University, Plymouth, United Kingdom [7]Josephine Bay Paul Center for Comparative Molecular Biology and Evolution, Woods Hole, MA, United States [8]Brown University, Providence, RI, United States

Marine plastic debris is a global conservation issue, raising concerns regarding ecological impacts. In the past, the focus of scientists was largely on macroplastic, whereas today it has shifted to microplastics. We examined the weight of evidence regarding perceived and demonstrated impacts of marine debris in general via a systematic review of the literature across 13 levels of biological organization (subatomic particle, atom, small molecule, macromolecule, molecular assemblage, organelle, cell, tissue, organ, organ system, organism, population, and assemblage). There were 366 perceived impacts across all levels of biological organization. Many were hypothesis-driven studies, wherein >83% were demonstrated impacts largely due to plastic debris. Overall, impacts were largely demonstrated at suborganismal levels of biological organization due to microdebris (<1 mm), while impacts at higher levels of organization (i.e., organism and above) were largely due to macrodebris (>1 mm). Decision-makers globally are requesting evidence of ecological harm to build effective policies. Here, we discuss the results and implication of our study with a focus on microplastic debris. In addition, we highlight some of the work on microplastic that has been published since completing our review (Rochmanet al., 2015).

While we agree that further information is needed to fill research gaps and provide assessments of ecological risk, our results suggest that there are several lines of evidence that plastic debris causes impacts across multiple levels of organization, including ecological.

REFERENCE

Rochman, C.M., Browne, M.A., Underwood, A.J., van Franeker, J.A., Thompson, R.C., Amaral-Zettler, L.A., 2015. The ecological impacts of marine debris: unraveling the demonstrated evidence from what is perceived. Ecology 97 (2).

Qualitative and Quantitative Investigations of Microplastics in Pelagic and Demersal Fish Species of the North and Baltic Sea Using Pyrolysis-GCMS: A Pilot Study

B.M. Scholz-Böttcher[1], M. Fischer[1], M.S. Meyer[1] and J. Gercken[2]

[1]Oldenburg University, Oldenburg, Germany [2]Institute for Applied Ecology (IfAÖ), Neu Broderstorf, Germany

Nowadays plastics of all shapes are omnipresent in the environment shown through several investigations over the last decade. Over the years of 6.11 Gt plastics produced since 1950 up to 10% are supposed to end up in the ocean as a final sink. Here, they are subjected to physical (fragmentation) and chemical transformations as well as several interactions with the bio- and chemosphere. With decreasing particle size their bioavailability increases.

The stomachs and intestines of pelagic and demersal fishes from distinct locations of the North and Baltic Sea were assessed for conspicuous particles microscopically and analyzed for their qualitative and quantitative small microplastics (S-MP) content regarding nine common polymers (PE, PP, PS, PVC, PET, PMMA, PC, PUR, PA) using pyrolysis-GCMS. The latter enables their qualitative and mass-related quantification at ng- to µg-trace level independent of any visual preselection. An enzymatic and chemical-oxidative cleanup as well as density separation were performed prior to analysis.

Particles >1 mm (except fibers) distinguishable as plastic were completely absent while over 45% of North- and Baltic Sea fish samples contained S-MP <1 mm. Fishes from Baltic Sea had a higher S-MP load and variety of polymers. All polymers were detected frequently apart from PP. There was no overall relationship between S-MP uptake, single detected polymer types and pelagic or benthic habitat of respective fish species. The amount and quality of S-MP seem to vary due to other factors that will be discussed. Overall, S-MP quantities are quiet low. S-MP contents below 20 µg per fish or pooled sample (3–6 fishes) for Baltic Sea and below 15 µg per pooled sample (2–5 fishes) for North Sea. Data were calculated based on a semi-quantitative estimation, since amounts of single polymer types are close to quantification limits.

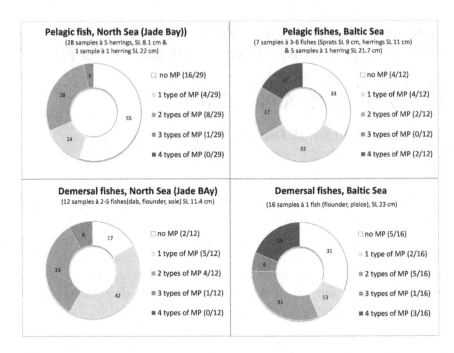

Relative frequencies of fish samples containing S-MP in the North (Jade Bay) and Baltic Sea and proportion of detected polymer types per sample.

Microplastics Extraction Methods for Small Fishes, on the Road to a Standard Monitoring Approach

B. Stjepan, L. Maiju and S. Outi

Finnish Environment Institute, Helsinki, Finland

Standardization of analytical methods is urgently needed for future steps in microplastics research in order to understand the impacts of microplastics on fish. Here, standard methods for microplastic extraction from the gastrointestinal tract (GIT) of fish were tested. In the laboratory, fish samples were spiked with high density-PE and PP particles made up of common household plastic items. Plastic was grinded and filtered to obtain two size classes for spiking: 1 mm 300 µm and 300 20 µm. In addition to particles, polyester fibers extracted from a fleece coat, and cotton fibers extracted from other clothing were used to spike the samples. Different protocols were tested, using both enzymatic and chemical digestion methods for both individual fish and pooled fish samples (multiple GIT of 5 to 10 fish).

Two common small species of fish were chosen as model organisms: Atlantic herring from the Baltic Sea (*Clupea harengus*) and European cisco (*Coregonus albula*) caught in Finnish freshwater lakes.

The whole GIT was isolated and treated in the digestion protocol. The gut contents were not extracted separately or visually examined to avoid contamination. As preliminary results of the research we propose a simple, cost effective, and applicable protocol for analyzing microplastics from small, common marine, and freshwater fishes on a larger geographical scale. One isolated GIT (weight 0.3–0.6 g) we treated with 10 mL NaOH concentration 1 mol/L and 5 mL of SDS concentration 5 g/L in the oven on 50°C for 2 h. Followed by vacuum filtration on Whatman glass microfiber filters (GF/F) pore size 0.6 μm or quantitative filter paper pore size 2–3 μm microplastic particles retained on filters were transferred in petri dish and observed under stereomicroscope.

Microplastic Effects in *Mullus surmuletus*: Ingestion and Induction of Detoxification Systems

C. Alomar[1], A. Sureda[2], M. Compa[1] and S. Deudero[1]

[1]Instituto Español de Oceanografía, Palma de Mallorca, Spain [2]University of Balearic Islands, Palma de Mallorca, Spain

Mullus surmuletus ($n = 415$) were sampled around the Balearic Islands to determine microplastic ingestion in gastrointestinal tracts and to analyze physiological alterations in liver. Fish samples were obtained from artisanal and trawling fishing boats in five commercial grounds. In laboratory, gastrointestinal tracts were examined under stereomicroscope for microplastic identification and 27% of the individuals had ingested microplastics with a mean value of 0.42 ± 0.82 microplastics/individual. Preliminary results show significant differences in microplastic ingestion between samples, with higher values found in the south area of Mallorca but without significant differences according to sex.

Further on, antioxidant enzymes—catalase and superoxide dismutase—and the detoxification enzyme glutathione *s*-transferase (GST) activities and malondialdehyde (MDA) levels as a marker of lipid peroxidation were determined in a subset of 40 samples of *M. surmuletus* with microplastics ($n = 20$) and without microplastics ($n = 20$). No significant differences were reported in the antioxidant enzymes activities and in MDA between both groups. The activity of GST was significantly increased in *M. surmuletus* containing microplastics ($p < 0.05$), suggesting that the presence of microplastics activates the detoxification mechanisms.

Assessment of Microplastics Present in Mussels Collected From the Scottish Coast

A.I. Catarino[1], V. Macchia[1], H. Barras[1], W. Sanderson[1],
R. Thompson[2] and T. Henry[1]

[1]Heriot-Watt University (HWU), Edinburgh, Scotland [2]Plymouth University (PU), Plymouth, United Kingdom

The presence of small (5 mm$-$1 μm) plastic debris, microplastics (MPs), in aquatic environments is recognized as among the highest international issues for environmental science and policy. MPs can be found in aquatic environments in suspension and/or associated with sediments and marine organisms can ingest MPs. Documentation of their presence and abundance in marine environments is an important initial step towards assessment of MPs effects on organisms. Particles have been found in most coastal areas where their presence has been investigated, but little is known about the abundance and types of MPs present on the Scottish coast. Our goal was to develop a standardized procedure to extract and quantify MPs in marine mussels, and then apply this method to assess MPs presence at specific locations along the Scottish coast. Complete *Mytilus edulis* soft tissue digestion was achieved with 1 M NaOH, 35% HNO_3 and by 0.1 UHb/mL protease, but use of HNO_3 caused destruction of nylon particles. The recovery of MPs spiked into mussels was similar ($93 \pm 10\%$) for NaOH and enzyme digestions. We recommend the use of industrial enzymes for soft tissue digestion since it is relatively easy, reproducible, provides good recovery rate of MPs from spiked tissue samples, and avoids the use of caustic chemicals that can damage polymers. This extraction method was applied to assess abundance of MPs (1) in mussels collected from various field sites on the coast of Scotland, and (2) in *M. edulis* deployed in purpose built cages together with passive samplers in an Edinburgh port (Port Edgar). Preliminary results of *M. edulis* individuals ($n = 18$) placed in cages in Port Edgar show a load of 2.0 ± 0.42 fibers, 0.2 ± 0.21 round shaped particles and 0.3 ± 0.59 films/g (wet weight of soft tissue), which are within the range of previous reports for *Mytilus* spp. from various European locations.

Bioavailability of Co-contaminants Sorbed to Microplastics in the Blue Mussel *Mytilus edulis*

A.I. Catarino, J. Measures and T.B. Henry
[1]Heriot-Watt University (HWU), Edinburgh, Scotland

Co-contaminants, such as polycyclic aromatic hydrocarbons (PAHs) and metals, can sorb to microplastics (MPs; 5 mm−1 μm) and desorption after their ingestion is of toxicological concern. The objective of this study was to assess the bioavailability of specific co-contaminants [cadmium (Cd^2), anthracene (ANT), and pyrene (PYR)] after they are sorbed to MPs. Polyvinyl chloride particles (125−250 μm) were exposed separately (180 rpm) for 24 h for ANT (100 μg/L) and PYR (20 μg/L) and 3 d for Cd^{2+} (50 μg/L) in MilliQ water. Particles were filtered and added at different concentrations (0−1 g/L) to beakers containing a mussel *Mytilus edulis* for evaluation of co-contaminant bioavailability. Concentration−response relationships were also determined for each co-contaminant in the aqueous phase based on changes in expression of biomarker genes in mussels. Gene expression was evaluated in gills and digestive gland with transcripts of superoxide dismutase (SOD) as a biomarker of PAH bioavailability and metallothionein-20 (MT-20) transcripts for Cd^{2+}. Gene expression was measured by quantitative reverse transcription PCR (RT-qPCR) after RNA extraction from tissues. Induction (fivefold relative to controls) of MT-20 in digestive gland of mussels exposed to MPs with sorbed Cd^{2+} indicated this metal was bioavailable. The expression of MT-20 in the digestive gland increased directly with particle concentration in a manner that was consistent with mussels exposed to Cd^{2+} in aqueous phase (c.a. 100 μg/L), but no significant changes in MT-20 expression were observed in the gills of mussel exposed to MPs with sorbed Cd^{2+}. Although SOD was significantly up-regulated upon mussel exposure to aqueous PAHs there was not a concentration relationship. The expression of SOD was induced in the digestive gland (mean 5.6 ± 4.66 SD and 6.4 ± 2.39 SD fold change in mussels exposed to PYR and ANT, respectively), but not in gills of mussels exposed to particles with sorbed PAHs.

Exploring the Effects of Microplastics on the Hepatopancreas Transcriptome of *Mytilus galloprovincialis*

M. Milan[1], C.G. Avio[2], S. Gorbi[2], M. Pauletto[1], M. Benedetti[2], G. d'Errico[2], D. Fattorini[2], T. Patarnello[1], F. Regoli[2] and L. Bargelloni[1]

[1]Università di Padova, Padova, Italy [2]Università Politecnica delle Marche, Ancona, Italy

Microplastics (MPs), represent a growing environmental concern for the oceans, notably for their potential of adsorbing chemical pollutants, thus representing a still unexplored source of exposure for aquatic organisms. Ingestion of these plastic particles has been demonstrated in a laboratory setting for a wide array of marine organisms. Bivalves, which filter large water volumes, are of particular interest since they may ingest huge quantities of particles exposing them directly to MPs present in the water column. Significant accumulation of plastic particles have been recently demonstrated in digestive tissues of mussels, causing notable histological changes with strong inflammatory responses, formation of granulocytomas, and lysosomal destabilization which increased with exposure time.

In the present study, to better elucidate pathways and molecular mechanisms of action of polystyrene MPs, individuals of the Mediterranean mussel *Mytilus galloprovincialis* were experimentally exposed to virgin polystyrene (PS) or pyrene-contaminated plastics (PS-PYR) for 7 days, and analyzed through a new species-specific Agilent oligo-DNA microarray platform representing 50,680 different contigs. Gene expression analyses revealed a total of 2143 and 1320 differentially expressed genes (DEGs) in response to PS and PS-PYR exposures, respectively. Functional annotation and enrichment analysis was then applied to DEGs to highlight the most significantly affected biological processes. Some of the most interesting enriched KEGG pathways/GO terms were lysosome, endosome, NOD-like receptor signaling pathway, response to bacterium, apoptosis, regulation of programmed cell death, citrate cycle, and arachidonic acid metabolism. In addition, significant changes were found in the modulation of genes involved in DNA repair, detoxification, and response to oxidative stress. Overall, both virgin and contaminated MPs induced several effects at the transcriptional level, highlighting the potential risk for organisms' health condition, especially under conditions of long-term, chronic exposure.

Characterization, Quantity, and Sorptive Properties of Microplastics Extracted From Cosmetics

I.E. Napper[1], A. Bakir[1,2], S.J. Rowland[2] and R.C. Thompson[1]
[1]Plymouth University, Plymouth, United Kingdom [2]Plymouth University, Plymouth, United Kingdom

Many cosmetics contain microplastics as exfoliating agents. Such products are an important source of microplastic contamination in the environment. This study characterizes, quantifies and then investigates the sorptive properties of plastic microbeads that are used as exfoliants in cosmetics. Polyethylene (PE) microbeads were extracted from several facial scrub products and were shown to have a wide size range (diameter $164-327\ \mu m$). We estimated that the products tested could each contain between 137,000 and 2,800,000 microparticles per bottle; therefore 4594−94,500 microbeads could be released in a single use (showing estimates per 1 mL). This could result in the UK population emitting up to 215 mg of PE/person/day; a total of 16−86 tons/year. To examine the potential for microbeads to accumulate and transport chemicals they were exposed to a binary mixture of 3H-phenanthrene and 14C-DDT in seawater. The potential for transport of sorbed chemicals by microbeads was broadly similar to that of polythene particles used in previous sorption studies. Microplastic size, rather than the average weight in each product, was found to be important as it had the greatest effect on abundance estimates. Their small size renders microplastics to be accessible to a wide range of organisms, and may facilitate the transfer of waterborne contaminants or pathogens. There are alternatives to the use of plastics as exfoliating particles, so they are a preventable source of microplastic contamination in the marine environment.

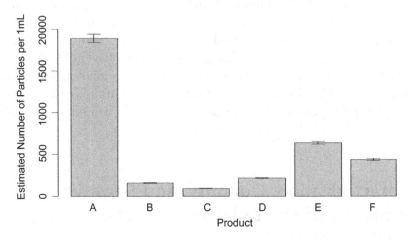

Estimates for the number of PE microbead particles in six brands of facial scrubs per 1 mL. Calculated using data from the volume weighted mean (n = 3, ±SD; correlating to the spread of the different amounts of particles calculated for high, medium, and low density PE).

Beach Sweep Initiatives on the Acadian Coastline in Atlantic Canada

O. Chouinard[1,2], C. Surette[1,2], B. Jorgensen[2,3] and J. Baztan[2,4]

[1]Université de Moncton, Moncton, NB, Canada [2]Marine Sciences For Society [3]Cornell University, Ithaca, NY, United States [4]Université de Versailles Saint-Quentin-en-Yvelines, Guyancourt, France

New Brunswick is a small rural province situated in Eastern Canada, bordering two different coastal areas, the Southern Gulf of Saint Lawrence and the Bay of Fundy. The Acadian coastline in the Southern Gulf of Saint Lawrence covers approximately 3500 km and is relatively pristine, compared to other regions of the world. Although microplastics do not seem to accumulate in great amount in this area, plastic debris are found and come mainly from activities such as fishing, tourism, and industrial activities. For example in the fisheries sector, lobster traps, fishing nets, buoys, and plastic ropes are regularly found on beaches. It is also common to find plastic bottles, bags or other objects used by tourists and citizens. For the past twenty years, Watershed associations and School Districts have partnered with the Beach Sweeps program to clean the Acadian Coast line of New

Brunswick. These Beach Sweeps activities are held each spring at the end May or beginning of June. In Southeastern New Brunswick, these activities are coordinated by Gestion H2O, a watershed group. Over the past years, eight other watershed associations, 27 schools, and many families or individuals have taken part in Beach Sweeps organized by Gestion H2O. Partnering with these local groups, who have substantial experience in monitoring the coastline and raising awareness to environmental issues, we can offer to act as a site in a global initiative to reduce plastic waste. We wish to meet with colleagues to discuss how to implement a global place-based observatory.

A Social-Ecological Approach to the Problem of Floating Plastics in the Mediterranean Sea

L.F. Ruiz-Orejón[1], R. Sardá[1] and J. Ramis-Pujol[2]
[1]Centre d'Estudis Avançats de Blanes (CEAB-CSIC), Blanes-Girona, Spain [2]Universitat Ramon Llull, Barcelona, Spain

Today, plastic pollution is a common problem in all oceans, but it is more severe in semienclosed systems such as the Mediterranean Sea. On 2009, the Fundación Innovación, Acción y Conocimiento (FIAyC) of Majorca (Balearic Islands, Spain) decided to replicate a part of the Mediterranean expeditions that Archduke Ludwig Salvator undertook one hundred years ago. So, the NIXE III project was initiated. Although the Archduke never saw plastic floating in the sea, the Pristine Sea that he saw serves as a baseline for the NIXE III observations. During our expeditions we are noting, sampling, and evaluating both the floating plastics and the background changes in the Mediterranean. The main objective of this talk is to provide all the new information generated about distribution, abundance, and size composition of floating plastics in the North-Western and Central Mediterranean Sea allocating them into the well-known DPSWR accounting framework (Driver-Pressure-State-Welfare-Response). In this way are aimed to organize the information related to this environmental problem and to link this with existing European policies.

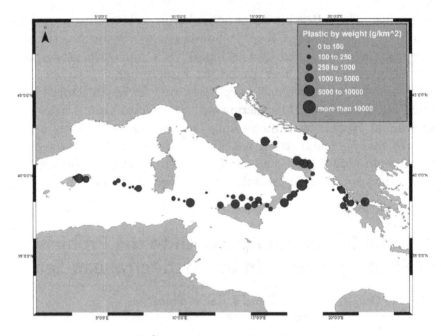

Plastic weight concentration (g dw/km^2) in the surface waters of the Mediterranean Sea.

News Splash? A Preliminary Review of Microplastics in the News

B. Jorgensen[1,2]
[1]Cornell University, Ithaca, NY, United States [2]Marine Sciences For Society

As cumulative scientific knowledge about microplastics continues its rapid increase within the scientific community, this media analysis was conducted to identify the news frames used to discuss microplastics in major world newspapers in order to better understand how this issue is presented to the general public through news media. This preliminary framing analysis of microplastics news coverage offers an overview of the general content focus and common frames used in articles on microplastics published in major global English-language newspapers over the past 10 years (January 1, 2006–April 1, 2016). Of the 197 articles found through LexisNexis, 87 mention microplastics casually, without providing a definition or substantive sentences about them, and were not included in the coding. The remaining 110 were coded for the presence/absence of standard frames (e.g., economic impact,

MICRO 2016. Fate and Impact of Microplastics in Marine Ecosystems.

human health, conflict, and attribution of responsibility) (deVreese, 2005) and emergent frames identified in the course of analysis (e.g., environmental conservation, science, policy, and risk/threat). Of these 110 articles:

- 36 focus on microbeads, and 32 of these do not mention other kinds of microplastic;
- 102 frame microplastics as "risks," and 50 use a "human health risk" frame; and
- 5 include a "conflict/disagreement" frame.

The findings also show that from 2006 to 2016, newspaper coverage of microplastics in marine environments has been steadily increasing, from two articles in 2006 to 30 in 2015, and already 16 articles in the first quarter of 2016. This increase seems correlated with the "Beat the Microbead" campaign, launched regionally in 2012 and internationally in 2013.

Future research should consider the impacts of such campaigns, and also investigate the approaches taken by researchers to share their work with news media.

Informing Policy Makers About State of Knowledge and Gaps on Microplastics in the Marine Environment

T. Bahri, U. Barg, E.G. Gamarro, P. Kershaw, J. Mendoza, J. Toppe and H. Savelli
Food and Agricultural Organization of the United Nations (FAO), Rome, Italy

For further information on this study, please contact the authors.

Microplastics in Cosmetics: Exploring Perceptions of Environmentalists, Beauticians, and Students

S. Pahl[1], J. Grose[1], A. Anderson[1], K. Wyles[2] and R. Thompson[1]
[1]Plymouth University (UoP), Plymouth, United Kingdom [2]Plymouth Marine Laboratory (PML), Plymouth, United Kingdom

Microplastics in the marine environment derive from larger plastic items breaking down ("secondary") and from particles originally manufactured at that size ("primary"). Cosmetics can contain such microplastics, for example in toothpaste, facial scrubs, etc. Society plays an important role in reducing microplastics and we need to understand perceptions of the wider public to move towards reductions. The present research explored awareness of microplastics in cosmetics in three groups: environmental activists, trainee beauticians, and university students in Southwest England. Focus groups were run using a standard protocol including the presentation of samples of plastic particles isolated from common high-street cosmetic products. Qualitative analysis showed that the issue of microplastics in cosmetics lacked visibility and immediacy for the beauticians and students, who were unaware of recent TV and press coverage on the topic.

However, the environmentalists were aware of the issue. Yet when participants were shown the amount of plastic in a range of familiar everyday cosmetics, they all expressed considerable surprise and concern at the quantities and potential impact. Regardless of the perceived level of harm, the consensus was that their use was unnatural and unnecessary. This research could inform future communication campaigns as well as industry and policy initiatives to phase out the use of microbeads.

Plastics and Zooplankton: What Do We Know?

P.K. Lindeque[1], A.W. McNeal[1] and M. Cole[1,2]
[1]Plymouth Marine Laboratory, Plymouth, United Kingdom [2]University of Exeter, Exeter, United Kingdom

For further information on this study, please contact the authors.

Microtrophic Project

A. Herrera, I. Martínez, T. Packard, M. Asensio and M. Gómez
Universidad de Las Palmas de Gran Canaria (EOMAR, ULPGC), Las Palmas de Gran Canaria, Spain

Microplastics, due to their small size, can be ingested by zooplankton and transferred throughout the marine food web. In addition to this physical hazard, there is a chemical hazard associated with microplastic ingestion. They concentrate persistent organic pollutants (POPs) at levels several orders of magnitude higher than those in the sea. POPs can be transferred to many organisms. They bioaccumulate and biomagnify, affecting higher levels in the food chain. In the Canary Islands, microplastics reach the coast via the south-flowing Canary Current. Due to the directionality of the predominant wind and currents, they are deposited mainly on the Islands' north coast. The first study carried out in the Canary Islands showed high levels of microplastic pollution, reaching 100 g/L of sediments on the north coast. This, plus the lack of information about the ecological impact of microplastic in this area prompted us to propose the "Microtrophic Project". We intend to find answers to several questions: How much microplastic is floating in the waters around the Canary Islands? Where are the microplastics accumulating? When are they more abundant? Are zooplankton and fish ingesting microplastics? The answers to these questions will help us to understand the potential impact of microplastics on marine biota. Finally, we propose laboratory investigations into the effects of microplastics and associated POPs on zooplankton physiology. We will use cultures of mysids, peracarida crustaceans with a highly relevant ecological role in Canary Island waters. They are the main food for many pelagic and coastal fish, here.

In addition to research, we support local awareness campaigns and beach clean ups while collaborating to disseminate information about the microplastic problem on our web and in Facebook. We also promote citizen science by involving students from the Lanzarote School of Fisheries and customers of Snorkelling Experience, an ecoturistic company, in our sampling program.

Source to Sink: Microplastics Ingested by Benthic Fauna From Discharge Points to Deep Basins in an Urban Model Fjord

M. Haave[1], T. Klunderud[1,2] and A. Goksøyr[2]
[1]Uni Research, Bergen, Norway [2]University of Bergen, Bergen, Norway

For further information on this study, please contact the authors.

Uptake and Toxicity of Methylmethacrylate-Based Nanoplastic Particles in Aquatic Organisms

A.M. Booth[1], B.H. Hansen[1], M. Frenzel[1], H. Johnsen[1] and D. Altin[2]
[1]SINTEF Materials and Chemistry, Trondheim, Norway [2]Biotrix, Trondheim, Norway

In the present study, the uptake and toxicity of two poly(methylmethacrylate)-based plastic nanoparticles (PNPs) with different surface chemistries (medium and hydrophobic) was assessed using aquatic organisms selected for their relevance based on the environmental behavior of the PNPs. Pure poly(methylmethacrylate) (medium; PMMA) PNPs and poly(methylmethacrylate-co-stearylmethacrylate) copolymer (hydrophobic; PMMA−PSMA) PNPs of 86−125 nm were synthesized using a mini emulsion polymerization method. Fluorescent analogs of each PNP (FPNPs) were also synthesized using monomer 7-[4-(trifluoromethyl)coumarin]acrylamide and studied. *Daphnia magna*, *Corophium volutator*, and *Vibrio fischeri* were employed in a series of standard acute ecotoxicity tests, being exposed to the PNPs at three different environmentally realistic concentrations (0.01, 0.1, and 1.0 mg/L) and a high concentration 500−1000 mg/L. In addition, sublethal effects of PNPs in *C. volutator* were determined using a sediment reburial test whilst the uptake and depuration of FPNPs was studied in *D. magna*. The PNPs and FPNPs did not exhibit any observable toxicity at concentrations up to 500−1000 mg/L in any of the tests except for PMMA–PSMA PNPs and FPNPs following 48 h exposure to *D. magna* (EC_{50} values of 879 and 887 mg/L, respectively). No significant differences were observed between labeled and nonlabeled PNPs, indicating the suitability of using fluorescent labeling for tracing of the NPs. Significant uptake and rapid excretion of the FPNPs was observed in *D. magna*.

Chemical structures of the two types of PNP, the fluorescent label, and the two types of FPNP synthesized and used in this study.

On the Potential Role of Phytoplankton Aggregates in Microplastic Sedimentation

M. Long, B. Moriceau, M. Gallinari, C. Lambert, A. Huvet, I. Paul-Pont, H. Hégaret and P. Soudant

Institut Universitaire Européen de la Mer (IUEM), Plouzané, France

For further information on this study, please contact the authors.

Hitchhiking Microorganisms on Microplastics in the Baltic Sea

S. Oberbeckmann[1], K. Kesy[1], B. Kreikemeyer[2] and M. Labrenz[1]

[1]Leibniz Institute for Baltic Sea Research Warnemuende (IOW), Rostock, Germany
[2]University of Rostock (UniR), Rostock, Germany

The project MikrOMIK aims at filling knowledge gaps regarding microplastics (MP) in the Baltic Sea. Project milestones are to determine (1) the distribution, sources, and sinks of MP in the Baltic Sea, (2) the role of MP as substrate for specific microbial populations, and (3) the health risks for the littoral states of the Baltic Sea emanating from MP. Here, we present the results of two microbiological experiments within the project. In an MP exposure experiment, polyethylene (PE), polystyrene (PS), and wooden pellets were incubated for 14 days at seven stations with various degrees of anthropogenic impact, located in the river Warnow (Rostock, Germany), the Baltic Sea, and a sewage treatment plant. Following incubation, the attached microbial communities were analyzed using high-throughput 16S amplicon sequencing (Illumina, MiSeq). The plastic and wood associated communities were compared to the free-living and particle-attached microbial communities from the respective background water. Generally, significant differences were found between free and attached-living communities, and between communities from the environmental stations (Warnow, Baltic) and the artificial system (treatment plant). No significant differences could be detected between the PE and PS associated communities overall, with solely few taxa showing a preference for one or the other polymer. The second experiment investigated the impact of the lugworm *Arenicola marina* on PS associated biofilms. Using a 16S rRNA gene based fingerprinting method, PS-attached communities were analyzed before and after passage through the worm. Both experiments revealed no specific enrichment of potentially pathogenic microorganisms on MP. The results of these experiments facilitate an understanding of the diversity, composition, and spatial variability of plastic associated microbial communities in the Baltic Sea. Upcoming analyses will comprise metagenomics in order to also learn about the function of these associated microbial communities.

Microplastic Prey? An Assay to Investigate Microplastic Uptake by Heterotrophic Nanoflagellates

A.M. Wieczorek, T.K. Doyle and P.L. Croot
National University of Ireland Galway, Galway, Ireland

With the recent increase of awareness of plastic waste in the ocean, plastics are reported in nearly every marine environment across the globe. Over time plastics are known to fragment into smaller particles through UV radiation and mechanical breakdown. Report a decrease in plastic particle size and an increase in global distribution over the past decades. Currently however, very little is known about the effects of the smallest plastic fragments (microspheres) on key marine biological processes. Small microspheres can have similar size ranges as ecologically important cyanobacteria such as *Synechococcus* ($\sim 0.7\,\mu$m). Thus, it was investigated whether heterotrophic nanoflagellate (HNF) grazers were able to distinguish between *Synechococcus* and microspheres of similar size as suggested by Sherr et al. (1987). If these plastic microspheres are ingested they have the potential to disrupt the base of the marine food web as they may cause physical harm as well as having toxic effects from plasticizers and sorbed toxins and heavy metals.

Interestingly microspheres have previously been used to investigate the feeding rates of such grazers. This research looks at the potential uptake of microspheres by HNF by utilizing a multiple approach using flow cytometry; via the phycoerythrin content of *Synechococcus*, Lysotracker staining of HNF and the SybrGreen staining of plankton/bacteria—combined in a dilution series to determine grazing and growth rates. By taking this approach we intend to gain insights into the grazing behavior of HNF on microspheres and photosynthetic bacteria.

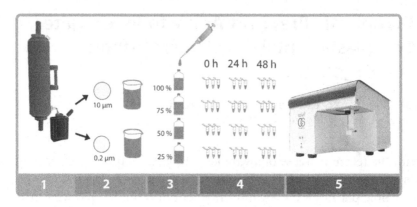

(1) Three water samples from the upper water column were taken during the SO245 Trans-Pacific cruise from Antofogasta, Chile, to Wellington, New Zealand, aboard the FS Sonne. (2) One half of each sample was filtered through a 10 µm filter in order to take out larger grazers and the other half through a 0.2 µm filter. (3) The 10 µm filtered water was then diluted with the 0.2 µm filtered water to give 100, 75, 50, and 25% concentrations. To each of these 20 µL of 0.73 µm microbeads were added. (4) Subsamples of the dilution series were taken after 0, 24, and 48 hours of incubation at natural conditions. (5) These samples were then run on the flow cytometer untreated, stained with SybrGreen and stained with Lysotracker.

Microplastics in Seafood: Identifying a Protocol for Their Extraction and Characterization

G. Duflos[1], A. Dehaut[1], A.-L. Cassone[2], L. Frère[2], L. Hermabessiere[1], C. Himber[1], E. Rinnert[3], G. Rivière[4], C. Lambert[2], P. Soudant[2], A. Huvet[5] and I. Paul-Pont[2]

[1]Agence nationale de sécurité sanitaire de l'alimentation, de l'environnement et du travail (Anses), Boulogne sur Mer, France [2]IFREMER – Institut Universitaire Européen de la Mer, Plouzané, France [3]IFREMER – Laboratoire Détection, Plouzané, France [4]Agence nationale de sécurité sanitaire de l'alimentation, de l'environnement et du travail (Anses), Maisons-Alfort, France [5]IFREMER – Laboratoire des Sciences de l'Environnement Marin (LEMAR), Plouzané, France

For further information on this study, please contact the authors.

Vast Quantities of Microplastics in Arctic Sea Ice—A Prime Temporary Sink for Plastic Litter and a Medium of Transport

M. Bergmann[1], I. Peeken[1], B. Beyer[1], T. Krumpen[1], S. Primpke[1], M.B. Tekman[1] and G. Gerdts[1]

[1]Alfred Wegener Institute, Helmholtz Centre for Polar and Marine Research (AWI), Bremerhaven, Germany

Although the Arctic covers 6% of our planet's surface and plays a key role in the Earth's climate it remains one of the least explored ecosystems. The global change induced decline of sea ice has led to increasing anthropogenic presence in the Arctic Ocean. Exploitation of its resources is already underway, and Arctic waters are likely important future shipping lanes as indicated by already increasing numbers of fishing vessels, cruise liners, and hydrocarbon prospecting in the area over the past decade. Global estimates of plastic entering the oceans currently exceed results based on empirical evidence by up to three orders of magnitude highlighting that we have not yet identified some of the major sinks of plastic in our oceans. Fragmentation into microplastics could explain part of the discrepancy. Indeed, microplastics were identified from numerous marine ecosystems globally, including the Arctic.

Here, we analyzed horizons of ice cores from the western and eastern Fram Strait by focal plane array based micro-Fourier transform infrared spectroscopy to assess if sea ice is a sink of microplastic. Ice cores were taken from land-locked and drifting sea ice to distinguish between local entrainment of microplastics *vs* long-distance transport. Mean concentrations of 2×10^6 particles/m^3 in pack ice and 6×10^5 particles/m^3 in land-locked ice were detected (numbers of fibers will soon be added). Eleven different polymer types were identified; polyethylene was the most abundant one. Preliminary results from four further ice cores from the central Arctic range in a similar order but the microplastics composition was very different. Calculation of drift trajectories by back-tracking of the ice floes sampled indicates multiple source areas, which explains the differences in the microplastic composition.

Preliminary analysis of snow samples taken from ice floes in the Fram Strait showed numerous fibers of yet unknown but most likely anthropogenic origin indicating atmospheric fallout as a possible pathway.

Our results exceed concentrations from the North Pacific by several orders of magnitudes. This can be explained partly by the process of ice formation, during which (organic) particles tend to concentrate by 1–2 orders of magnitude compared with ambient seawater. However, the magnitude of the difference indicates that Arctic sea ice is a temporal sink for microplastics. Increasing quantities of small plastic litter items on the seafloor nearby, which is located in the marginal ice zone corroborate the notion that melting sea ice releases entrained plastic particles and that sea ice acts as a vector of transport both horizontally and vertically to underlying ecosystem compartments.

Microplastics in the Bay of Brest (Brittany, France): Composition, Abundance, and Spatial Distribution

L. Frère[1], I. Paul-Pont[1], C. Lambert[1], E. Rinnert[2], C. Lacroix[3], P. Soudant[1] and A. Huvet[1]

[1]IUEM, Plouzané, France [2]Ifremer, Plouzané, France [3]CEDRE, Brest, France

World production of plastics has increased from 1.7 to 311 million tons between 1950 and 2014. This led to a major contamination of the worldwide marine ecosystems, recently estimated at more than 5 trillion plastic pieces floating the surface of the oceans. Microplastics (MP, <5 mm) are introduced into aquatic environments both directly in industrial raw material (plastic pellets, cosmetics, and clothing) and from the fragmentation of larger plastics into MP. At the European level, the Marine Strategy Framework Directive has defined marine litter as a full descriptor of the marine environment with a focus on MP and degradation products as a main issue. The present study focused on the composition and characteristics of microparticles collected, and their spatial distribution within surface water of the Bay of Brest (Brittany, France). Distinct environments were sampled to estimate the impact of the main sources of urban, industrial, harbor, and agricultural perturbations (3 stations per area): north area (highly urbanized zone), central area (oceanic zone with high levels of water mixing), and south part (estuary influenced by important shellfish farming activities) and in two matrices (surface water and sediment). Each collected particle was analyzed using a combination of static image analysis of particles and automated Raman microspectroscopy allowing the counting, size, shape, and chemical composition of a large number of particles. For surface water sampling, all stations contained MP with a mean abundance of 0.296 ± 0.459 MP/m^3.

Plastic Litter: A New Habitat for Marine Microbial Communities

C. Dussud[1], M. Pujo-Pay[1], P. Conan[1], O. Crispi[1], A. Elineau[1], S. Petit[1], G. Gorsky[1], M.-L. Pedrotti[1], P. Fabre[2], M. George[2] and J.-F. Ghiglione[1]

[1]Université Pierre et Marie Curie (UPMC), Paris, France [2]Website Université Montpellier, Montpellier, France

Plastic litter has become the most common form of marine debris and a major and growing global pollution concern. In marine waters, plastic fragments are rapidly colonized by microorganisms, characterized by a very diverse community called "plastisphere." The research on biodegradability of plastics began in the early 1980s and numerous papers provide culture-based evidence of various microorganisms able to degrade a variety of plastics under controlled conditions.

Our researches focus on the characterization of the microbial communities living at the surface of microplastics (<5 mm). Most of our samples were collected during the Tara Mediterranean expedition, which demonstrated that 100% of the sea surface of Mediterranean Sea (from coast to open ocean) is contaminated by microplastics. We evaluated the abundance, activity, and diversity of the microbial communities using a wide range of techniques, including optic and atomic force microscopy, flow cytometry, prokaryotic heterotrophic production, ectoenzymatic activity, and metagenomics analysis.

Other studies performed under microcosm conditions (2L) but directly connected to the sea allowed us to follow the colonization during 4 months of new selected plastics, including low density polyethylene, oxo-biodegradable and biosourced polymers together with artificially aged plastics.

All our analyses were coupled with chemical characterization of plastics and showed that plastispheres were different depending on the type of plastic. Our results indicated also that microbes living in microplastics present distinct patterns from surrounding free-living and particle-attached fractions in seawater, implying that plastic serves as a novel ecological habitat in the ocean.

The Effects of Microplastic on Freshwater *Hydra attenuatta* Morphology and Feeding

F. Murphy, L. Prades, C. Ewins and B. Quinn
University of the West of Scotland, Paisley, Scotland

Microplastic pollution has been a growing concern in the aquatic environment for a number of years. It's abundance in the environment has invariably lead them to come into contact with a variety of different aquatic species. With many capable of ingesting these plastics, the impact of which is not fully understood. The majority of research studying microplastic pollution has focused upon the marine environment and species with little attention paid to the freshwater. Here we examine the effect of microplastics on a freshwater cnidarian, *Hydra attenuata*. *Hydra* play a vital role in the planktonic make up of slow moving water bodies which they inhabit and are sensitive environmental indicators. *Hydra* were exposed to polyethylene particles (<400 μm) extracted from facewash at different concentrations (Control, 0.01, 0.02, 0.04, 0.08 g/mL). The ingestion rate of prey (*Artemia*) and microplastics were recorded every 30 min for 120 min. The preliminary results of this study show that *H. attenuata* are capable of ingesting microplastics. After 60 min of exposure the *Hydra* ingested a lower amount of *Artemia* in all concentrations compared to the control. This became even more evident at the 90 and 120 min mark. No *Artemia* were ingested at the highest concentration used (0.08 g/mL) after the 90 min mark. This shows that the presence of microplastics can interfere with the feeding of *Hydra*, either by reducing the amount of prey ingested or by preventing feeding completely. Exposure to the microplastics caused significant changes to the morphology of the *Hydra*, however these changes were nonlethal. This study demonstrates that freshwater *Hydra* are cable of ingesting microplastics and that microplastic can impact the feeding of freshwater organisms.

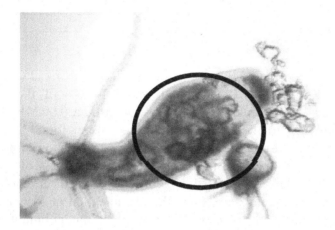

Photo of Hydra *with ingested microplastic.*

Evidence of Microplastic Ingestion in Elasmobranchs in the Western Mediterranean Sea

S. Deudero, A. Frank and C. Alomar

Instituto Español de Oceanografía, Palma de Mallorca, Spain

A total of 146 elasmobranchii were analysed for microplastic (<5 mm) identification in gastrointestinal tracts. Two sampling locations (Palma and Soller) were sampled around Mallorca Island (W Med). Four different species (*Galeus melastomus*, *Etmopterus spinax*, *Scyliorhinus canicula* and *Chimaera monstrosa*) were captured with commercial trawling boats at 600 m depth (Table 1). In general terms, 23.08% of the sampled elasmobranchii in Soller had ingested microplastics (MPs) whereas only 13.83% of all sampled elasmobranchii in Palma showed MPs in their gastrointestinal tracts. According to *G. melastomus*, which was the most abundant species ($n = 131$), 18.18% of the individuals

Table 1 Sampled Species, Total Number of Individuals and Biological Parameters (Size Range, Fresh Weight Range, and Ratio of Males (M) to Females (F)					
Species	Number of Individuals		Sizer	Fresh Weight Range (g)	Ratio
Galeus melastomus	131	115	–	5–400	1:0.6
Etmopterus spinax	8	165		21–122	1:4
Scyliorhinus canicula	5	370	–	162–236	1:4
Chimaera monstrosa	2	210		13	–
Total	146				

captured in Soller had MPs in their gastrointestinal tracts with mean values of 0.32 ± 0.11 MPs/individual, while 14.94% of the samples from Palma showed microplastic ingestion (0.32 ± 0.09 MPs/individual), however, no significant differences were found ($p > 0.05$, PERMANOVA). In addition, microplastic ingestion showed no variation among individual size of *G. melastomus*, regardless of the wide range of sampled sizes (115–560 cm).The studied elasmobranchs exhibit microplastic ingestion, thus the identification of MPs in digestive contents indirectly reflects the presence of this pollutants in the demersal habitats of the marine environment.

Do Microplastic Particles Impair the Performance of Marine Deposit and Filter Feeding Invertebrates? Results From a Globally Replicated Study

M. Lenz and M. Wahl
GEOMAR Helmholtz Centre for Ocean Research Kiel (GEOMAR), Kiel, Germany

For further information on this study, please contact the authors.

Occurrence of Potential Microplastics in Commercial Fish From an Estuarine Environment: Preliminary Results

F. Bessa[1], P. Barria[1], J. Neto[1], J. Marques[1] and P. Sobral[2]

[1]Universidade de Coimbra, Coimbra, Portugal [2]Universidade Nova de Lisboa, Monte de Caparica, Portugal

The ingestion of microplastic particles by marine biota has been reported for a variety of different species, however, data on abundance in natural populations is still limited. In this study, we assessed the presence of anthropogenic debris and potential microplastics in the gastrointestinal tracts of four commercialized species of fish: *Diplodus vulgaris* (demersal), *Solea solea* (benthic), *Dicentrarchus labrax* (demersal), and *Platichthys flesus* (benthic) from the Mondego estuary, in Portugal. Samples were collected along the estuary in two areas (the most downstream station in the mouth of the estuary and another located more upstream) during June and October 2014. For the moment, a total of 90 individuals were analyzed and ingested anthropogenic particles were found in 42 of the digestive tracts of fish. These particles were mainly fibers and fragments of various sizes and colors. There was no evident spatial pattern of particles distribution in the fishes in the estuary but a higher frequency of ingested particles was recorded for the common two-banded sea bream (*D. vulgaris*) with 80% of fish having anthropogenic particles in their digestive tract. To confirm if those particles are plastic, selected samples will be analyzed by means of Fourier transform infrared spectroscopy determining the polymer types.

These preliminary results highlight the need for further studies that can more fully evaluate the environmental impacts of plastic ingestion by fishes in natural environments, with particular attention to those directly used for human consumption.

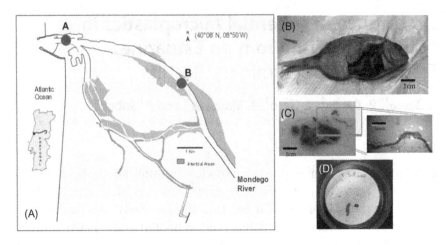

(a) The Mondego estuary with the location of the sampling stations (A and B); (b) a common two-banded sea bream dissected; (c) and (d) selected anthropogenic particles found in the digestive tract of fish.

Improvements and Needs of Microplastics Analytical Control at Open Sea Opportunities for Monitoring at Canary Islands

D. Vega-Moreno[1], M.Dolores. Gelado-Caballero[1], C. Barrera[2] and J.J. Hernández-Brito[2]

[1]University of Las Palmas de Gran Canaria (ULPGC), Las Palmas de Gran Canaria, Spain
[2]Oceanic Platform of the Canary Islands (PLOCAN), Telde, Spain

Most microplastics collected at beaches and oceanic regions show high erosion, with rounded shapes. This can involve, as product of this erosion, the generation of "plastic dust" or "plastic powder" much thinner, with size between several nm to $100-200\,\mu m$, undetectable by established microplastic extraction methods (lower size limit $330\,\mu m$). This plastic powder could be represented an important amount over total plastic presents into the oceans. There are big differences between estimated total plastic data at oceans and sampled data, without explanation yet (Jambeck et al., 2015; Nerland et al., 2014). Called small microplastics or nanoplastics, it is necessary to improve the studies about them, because they could represent a bigger problem than microplastics despite they are not visible to the naked eyed.

The main reasons are:

- High surface/volume relation (one or two magnitude order higher than microplastics). Related with high adsorption capacity of POPs (Persistance Organic Pollutants) and PBTs (Persistance, Bioacumulative and Toxic substances) (Mattsson et al., 2015).
- Nano-sized particles appearing to be taken up more easily by organisms and being more difficult to excrete than the larger particles (Manabe et al., 2011). Due to their little size, nanoplastics are incorporated to intracellular level at organisms that ingest them (GESAMP, 2015).
- Nanoplastics can represent a problem at toxicological level due to POPs, PBTs but also by chemical substances which are formed, that can act as endocrine disruptors through food chain, increasing the effect by bioaccumulation and biomagnification.

For nanoplastics sampling it will be necessary to establish a new extraction method by high volume sample filtration, with equipment that could be similar to those used by phytoplankton analysis. It is also necessary to select new analytical methodologies for the determination and identification of these compounds, because used methods at microplastics (FT-IR and Raman spectroscopy) are limited by low amount of sample or lower particle sizes. Indirect determination methods can be an option for quantification of microplastics smallest fraction, based at the identification of plastic components and toxic substances associated to them at trace level. Canary Islands, and specifically Gran Canaria Island, gather the institutional, environmental and scientific conditions to the establishment of microplastics monitoring network at oceanic level in North Atlantic, in collaboration between the University of Las Palmas de Gran Canaria and the Oceanic Platform of the Canary Islands (PLOCAN). Moreover, this region and network gather also the potential to arrange a new project where lay the groundwork for the extraction, monitorization and analytical control for nanoplastics determination at open sea.

REFERENCES

(GESAMP)-Joint Group of Expert on the scientific aspect of marine environmental protection. 2015. Source, fate and effects of microplastics in the marine environment: a global assessment.

Jambeck, J.R., Geyer, R., Wilcox, C., Siegler, T.R., Perryman, M., Andrady, A., et al., 2015. Plastic waste inputs from land into the ocean. Science 347 (6223), 768–771.

Manabe, M., Tatarazako, N., Kinoshita, M., 2011. Uptake, excretion and toxicity of nano-sized latex particles on medaka (Oryzias latipes) embryos and larvae. Aquat. Toxicol. 105 (3-4), 576–581. Available from: http://dx.doi.org/10.1016/j.aquatox.2011.08.020.

Mattsson, K., Hansson, L.-A., Cedervall, T., 2015. Nano-plastics in the aquatic environment. Environ. Sci.: Processes Impacts 17 (10), 1712–1721. Available from: http://dx.doi.org/10.1039/C5EM00227C.

Nerland, I.L., Halsband, C., Allan, I., Thomas, K.V., 2014. Microplastics in marine environments: Occurrence, distribution and effects, Norwegian Institute for Water Research, REPORT SNO. 6754-2014.

A Novel Method for Preparing Microplastic Fibers

M. Cole
University of Exeter, Exeter, United Kingdom

Microscopic plastic (microplastic, <5 mm) debris is a widespread pollutant impinging upon freshwater and marine ecosystems across the globe. The risks microplastics pose to aquatic life arse of environmental concern. Laboratory experiments have revealed a range of animals can ingest microplastic, with repercussions for their feeding, growth, and reproduction. Microplastics sampled from the water surface, sediments, and animal tissues are dominated by irregular fragments and fibers. Conversely, laboratory studies predominantly use spherical microbeads as representative microplastics, since they are easy to image and source. While spherical microplastics can clearly elicit adverse health responses, fibrous microplastics may pose an even greater threat. Nanoscopic and microscopic fibrous materials (e.g., asbestos fibrils, carbon nanotubes) can result in carcinogenesis and fibrosis, whereas particles of the same material in particulate form are often benign. The aim of this project was to develop a method for generating microfibers of predetermined polymer type, shape, and size to enable laboratory studies of bioavailability and impact. Preparation of microfibers using a novel "cryotome" method is described, and the resultant microplastics compared with the outputs from alternate methods using electron microscopy and image analysis. It is anticipated that these results will prompt a rethink of future experimental design and provide researchers with the best methods for incorporating fibrous microplastics into their exposures, allowing scientists to develop a greater understanding of the bioaccumulation and biological effects of microplastics in the oceans.

A New Approach in Separating Microplastics From Freshwater Suspended Particulate Matter

S. Hatzky, C. Kochleus, E. Breuninger, S. Buchinger, N. Brennholt and G. Reifferscheid
German Institute for Hydrology, Koblenz, Germany

For further information on this study, please contact the authors.

Microplastics—Microalgae: An Interaction Dependent on Polymer Type

F. Lagarde[1] and A. Caruso[2]
[1]University of Le Mans, Le Mans, France [2]Université du Maine, Le Mans, France

Recent studies showed that milli- and microplastic fragments sampled from the environment constitute a new habitat for a large diversity of organisms both in oceanic and freshwater systems. This important colonization of plastics can play a significant role in their buoyancy as biofouling can lead to an increase of the polymer fragments density. But the changes in polymer densities appear strongly variable depending upon seasons, colonizing species, and polymer type. Laboratory experiments can help to understand the changes in polymer densities during colonization.

In this study, the interactions between polypropylene (PP) and high density polyethylene (HDPE) microfragments, and a model freshwater microalgae were investigated. The presence of high concentrations of plastic fragments did not directly impact the growth of *Chlamydomonas reinhardtii* microalgae and the expression of three genes involved in the stress response (two genes of oxidative stress and one of apoptose) was studied in real time PCR. The expression of these genes was not modified after contact with microplastics. In the first days of contact, the microalgal colonization on the two polymers was observed by microscopy, infrared spectroscopy, and through water contact angle measurements. Both polymers exhibited a similar pathway of colonization. However, after 20 days of contact, in the case of PP only, hetero-aggregates constituted of microalgae, microplastics, and exopolysaccharides were formed with a final density close to 1.2. Such hetero-aggregates can be one of the ways for the vertical transport of PP microplastics to sediments. Besides, after more than 70 days of contact with microplastics, the microalgae genes involved in the sugar biosynthesis pathways were strongly overexpressed, particularly in the case of HDPE. This work presents first evidence that depending on their chemical nature, microplastics will follow different fate in the environment.

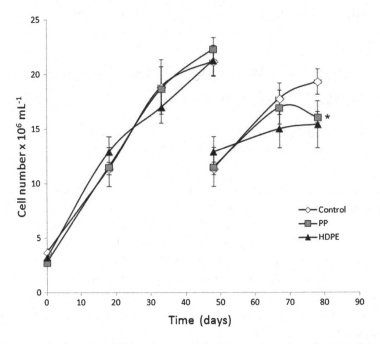

Cell density: total cell number of Chlamydomonas reinhardtii microalgae submitted to control conditions (***), 100 mg of PP (***) or HDPE (▲) fragments versus time of contact. Vertical bars represent S.E. values (n = 3). Means were compared between control and plastic conditions at each date of measurement; (*) significant differences at $p < 0.05$ (Student test).

Nearshore Circulation in the Confital Bay: Implications on Marine Debris Transport and Deposition at Las Canteras Beach

L. Mcknight[1,2], M. Rodrigues[3], A.B. Fortunato[3], G. Clarindo[1,2] and G. Rodríguez[1,2]

[1]University of Las Palmas de Gran Canaria (ULPGC), Las Palmas de G.C., Spain
[2]Institute of Environmental and Natural Resources Study (i-UNAT), ULPGC, Las Palmas de G.C., Spain [3]National Laboratory for Civil Engineering, Lisbon, Portugal

Preliminary results of a study on the surface circulation in the Confital Bay, and particularly in the nearshore zone of Las Canteras beach (Gran Canaria, Spain), are presented. Results are derived by means of experimental measurements of tidal elevations, wind, wind-generated waves, and Eulerian and Langrangian current measurements, as well as an advanced numerical model of coastal circulation. Inputs to the numerical model are bathymetric, tidal, and wind conditions. Model performance is examined by checking Eulerian and Lagrangian circulation patterns predicted by the model against empirical observations. Results are used to explore implications of surface circulation wind, tides, waves, and currents conditions within the bay on marine debris transport and deposition on Las Canteras beach, an urban beach with high density of users, with the particular geomorphological characteristic of being a semienclosed beach by a natural rocky reef, and an ecological protected area of international interest.

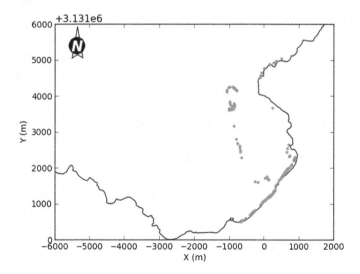

Illustrates instantaneous surface particle tracking (time: 199.58 h) released along the beach at depth of 2.20 m, under stationary wind conditions coming from NW.

New Approaches for the Extraction and Identification of Microplastics From Marine Sediment

M. Kedzierski[1], V. Le Tilly[1], P. Bourseau[1,2], G. César[1], O. Sire[1] and S. Bruzaud[1]

[1]Université Bretagne Sud, Lorient, France [2]Université de Nantes, Saint-Nazaire, France

For further information on this study, please contact the authors.

Using Physical and Chemical Characteristics of Floating Microplastics to Investigate Their Weathering History

K.L. Law[1], J. Donohue[1], T. Collins[1], K. Pavlekovsky[1], A. Andrady[2] and G. Proskurowski[3,4]

[1]Sea Education Association, Woods Hole, MA, United States [2]North Carolina State University, Raleigh, NC, United States [3]University of Washington, Seattle, WA, United States [4]MarqMetrix, Seattle, WA, United States

Microplastics collected from the open ocean offer few clues about their origin and history. Presently there is no method to determine how long ocean plastic has undergone environmental weathering, how quickly fragmentation has occurred, or how small microplastic particles will ultimately become before (or if) they are fully degraded by microbial action. We meticulously examined physical and chemical characteristics of open ocean microplastic particles collected over a 16-year period for clues about their weathering history.

Nearly 3000 microplastic particles collected in the western North Atlantic and eastern North Pacific between 1991 and 2014 were analyzed to determine polymer type, material density, mass and particle size, and were used to create a detailed catalogue of common microscopic surface features likely related to environmental exposure and weathering. Polyethylene and polypropylene can typically be distinguished by visual microscopy alone, and their particular characteristics lead us to hypothesize that these two resins weaken and fragment in different ways and on different time scales.

We hypothesize that regional differences in average or median particle mass and size are a relative indicator of age (time of exposure), where accumulation zones that retain particles for long periods of time have statistically smaller fragments compared to regions closer to presumed sources. Differences in particle form (i.e., fragment, pellet, foam, line/fiber, film) might also reflect proximity to sources, as well as form-dependent removal mechanisms such as density increase and sinking (Ryan, 2015). Changes in particle composition over time in subtropical gyre reservoirs could provide clues about changes in input as well as mechanisms and time scale of removal. Finally, a subset of resin pellets collected at sea were also analyzed using FTIR-ATR and/or FTIR microscopy for signatures of chemical degradation (e.g., carbonyl index) that are related to physical weathering characteristics such as color, quantified by the yellowness index.

REFERENCE

Ryan, P., 2015. Does size and buoyancy affect the long-distance transport of floating debris? Environ. Res. Lett. 10, 084019.

The Size Spectrum as Tool for Analyzing Marine Plastic Pollution

E. Martí[1], C.M. Duarte[2] and A. Cózar[1]
[1]Universidad de Cádiz, Puerto Real, Spain [2]King Abdullah University of Science and Technology, Thuwal, Kingdom of Saudi Arabia

Marine plastic debris spans over six orders of magnitude in lineal size, from microns to meters. The broad range of plastic sizes mainly arises from the continuous photodegradation and fragmentation affecting the plastic objects. Interestingly, this time-dependent process links, to some degree, the size to the age of the debris. The variety of plastic sizes gives the possibility to marine biota to interact and possible take up microplastics through numerous pathways. Physical processes such as sinking and wind-induced transport or the chemical adsorption of contaminants are also closely related to the size and shape of the plastic items. Likewise, available sampling techniques should be considered as partial views of the marine plastic size range. This being so and given that the size is one of the most easily measurable plastic traits, the size spectrum appears as an ideal frame to arrange, integrate, and analyze plastic data of diverse nature. In this work, we examined tens of thousands of plastic items sampled from across the world with the aim of (1) developing and standardizing the size-spectrum tool to study marine plastics, and (2) assembling a global plastic size spectrum (GPSS) database, relating individual size measurements to abundance, color (129 tons), polymer type, and category (rigid fragments, films, threads, foam, pellets, and microbeads). Using GPSS database, we show for instance the dependence of plastic composition on the item size, with high diversity of categories for items larger than 1 cm and a clear dominance ($\sim 90\%$) of hard fragments below, except for the size interval corresponding to microbeads (around 0.5 mm). GPSS database depicts a comprehensive size-based framework for analyzing the marine plastic pollution, enabling the comparison of size-related studies or the testing of hypothesis.

Automated Analysis of μFTIR Imaging Data for Microplastic Samples

S. Primpke[1], C. Lorenz[1], R. Rascher-Friesenhausen[2,3] and G. Gerdts[1]

[1]Alfred-Wegener-Institute Helmholtz Centre for Polar and Marine Research, Helgoland, Germany
[2]Hochschule Bremerhaven, Bremerhaven, Germany [3]Fraunhofer MEVIS, Bremen, Germany

The pollution of the oceans with plastic particles smaller than 5 mm, called microplastics is moving into the focus of science and governments. To determine the amount of microplastics several steps are necessary, starting with the sampling, work up, and finally analysis. Each step has its own challenges due to small size of the particles. For analysis the imaging with μFTIR microscopy is a powerful tool allowing the analysis of complete filters. Systematic screening for optimal conditions and filter materials have already been performed. This method has a high time demand regarding the measurement and data interpretation. While the measurement is performed mostly by the spectrometer, the interpretation has to be made by hand on the basis of false color images. To overcome the manual part we developed a novel approach based on the BrukerOPUS1 Software to decrease the high time demand for the analysis of microplastics. With this approach it was possible to analyze measurement files from focal plane array (FPA) FTIR mapping containing up to 1.8 million single spectra. These spectra were compared with a database of different synthetic and natural polymers by various methods. By benchmark tests their performance was monitored with the focus on accuracy and data quality. After optimization high quality data was generated, which allowed image analysis. Based on these results an approach for image analysis was developed, giving information for the particle size distribution for each polymer type, particle distribution on the filter and polymer distribution for the particles. It was possible to collect all data with relative ease even for complex sample matrices. This approach has significantly decreased the time demand for the interpretation of FTIR-imaging data and increased the generated data quality.

Solar Radiation Induced Degradation of Common Plastics Under Marine Exposure Conditions

A.L. Andrady[1], K.L. Law[2], J. Donohue[2] and G. Proskurowski[3]
[1]North Carolina State University, Raleigh, NC, United States [2]Sea Education Association, Woods Hole, MA, United States [3]University of Washington, Seattle, WA, United States

For further information on this study, please contact the authors.

Using the FlowCam to Validate an Enzymatic Digestion Protocol Applied to Assess the Occurrence of Microplastics in the Southern North Sea

C. Lorenz[1], L. Speidel[1], S. Primpke[1] and G. Gerdts[1]

[1]Alfred Wegener Institute Helmholtz Centre for Polar and Marine Research (AWI), Helgoland, Germany

As the plastic production has been rising since the last five decades, so does the concern for the occurrence of microplastic particles (<5 mm) in the marine environment during the past years. But still by now the extent of this microplastic pollution of coastal waters and the open ocean remains unclear. Since monitoring the abundance of microplastics in the marine environment is requested by the Marine Strategy Framework Directive (MSFD) standardized and reliable methods for the detection of microplastics are urgently needed.

Studies differ mainly in their purification methods, aiming to reduce biogenic material in environmental samples without altering the plastic polymers to facilitate a clear assignment of the microplastics. In the present and ongoing study the purification method consists of a treatment with technical enzymes and detergents to reduce the use of oxidants and avoid the use of strong acids as well as the subsequent identification and quantification of microplastics applying Focal Plane Array (FPA)-based μFourier Transform Infrared (μFT-IR) spectroscopy.

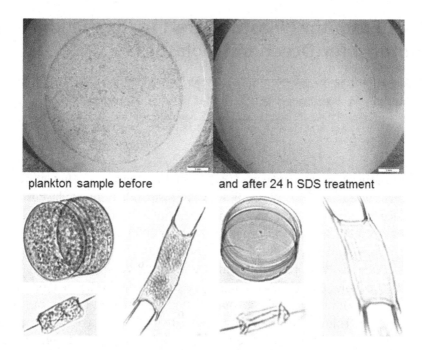

plankton sample before   and after 24 h SDS treatment

A new approach involving the FlowCam (Fluid Imaging Technologies) was used to validate the efficacy of the applied digestion protocol. Hereby untreated plankton samples were run through the FlowCam and compared to the samples after treatment with the various technical enzymes (protease, cellulase, and chitinase), sodium dodecyl sulfate and hydrogen peroxide according to protocol regarding changes to particle count, particle sizes, and appearance. Furthermore a metabolic fingerprint of every digestion step was drawn up applying the same μFT-IR spectroscopic analysis used for the identification of microplastics. Exemplary results derived by applying the validated protocol show the occurrence of microplastics in sediments of the German coastal waters and highlight the need for a further assessment of the microplastic pollution in various compartments i.e., sediments and surface waters and investigating the link between these two.

DNA From the "Plastisphere": An Extraction Protocol for Ocean Microplastics

P. Debeljak[1], M. Proietti[2], J. Reisser[1], M. Pinto[3], C. Baranyi[3], B. Abbas[4], M. Van Loosdrecht[4], B. Slat[1] and G. Herndl[1,3]

[1]The Ocean Cleanup (TOC), ES Delft, Netherlands [2]Universidade Federal do Rio Grande, Rio Grande, Brazil [3]University of Vienna, Vienna, Austria [4]Delft University of Technology (TU Delft), BC Delft, Netherlands

The ubiquity of microplastics (<5 mm) in oceans worldwide raises concern about their ecological effects. Suspended microplastics are ingested by marine wildlife and may accumulate in the food web along with associated toxins and organisms which colonize plastic. Owing to their widespread dispersal, plastic debris items also hold the potential to function as vectors for harmful and invasive species. Currently, genetic techniques are increasingly used to gain better insight into epiplastic communities, which may include pathogens and organisms that play a role in the fate of plastics (e.g., hydrocarbon-degrading bacteria and fungi). Despite the growing application of molecular techniques to study ocean microplastic colonizers, to date there is no comparative study on the most efficient way to extract DNA from biofilm found on microplastics. In this context, the present study compared the quantity of DNA obtained via five extraction methods (four commonly used extraction kits, and the standard phenol:chlorophorm purification) and two mechanical lysis techniques (microbeading and cryogenic grinding with liquid nitrogen). These were applied to four quantities of microplastics (1, 15, 50, and 80 fragments per extraction) of two size classes (0.05–0.15 and 0.15–0.5 cm) collected from the North Pacific accumulation zone. This study will provide a detailed protocol for extracting DNA from marine microplastics, considering different sizes and amounts of microplastics, suitable for further molecular analysis, such as for next generation sequencing.

Quality Assurance in Microplastic Detection

C. Wesch, R. Klein and M. Paulus

Trier University, Trier, Germany

Monitoring the occurrence of microplastics in environmental samples is challenging as false identification of microfibers as microplastics, airborne contamination of samples, nonstandardized procedures, and insufficiently used analytical techniques make any comparison of the various microplastic studies difficult. Thus, misidentification or overestimation of microplastics in environmental samples cannot be avoided and presumably the exposure of organisms as well as aquatic ecosystems to microplastics is frequently overestimated. Therefore, in order to achieve a high degree of quality assurance in the future and to ensure reproducibility and representativeness of microplastic results as well as to facilitate global comparisons, the selection of suitable techniques and applied analytical methods should be standardized. This can be achieved by Standard Operating Procedures, which describe methodologies for basic environmental monitoring processes and are developed to assure a high degree of standardization. Consequently, results of microplastic monitoring studies should only be assumed, if they are backed up by quality assurance measures including the documentation of all monitoring steps. Accordingly, our present study highlights the necessity for standardized measures and gives recommendations for future practical procedures in microplastic detection: at first, looking at the most recent developments of methods for the identification of microfibers, we urgently recommend an adoption of spectroscopic techniques in standardized microplastic monitoring schemes to avoid any misidentification of natural for synthetic microfibers. Additionally, we supply a proved approach to prevent airborne fibers contamination of environmental samples under standardized clean air conditions. Designed for the protection of biological specimens, the use of a clean bench was tested and is recommended by our results for microplastic laboratory application in future research as any airborne contamination could be excluded.

PART *II*

Abstracts From Posters

Are the Sinking Velocities of Microplastics Altered Following Interactions With *Austrominius modestus* and Sediment Particles?

R. Adams and R. Thompson

Plymouth University, Plymouth, United Kingdom

The ever-increasing presence of plastic debris in the marine environment has become a globally recognised environmental concern. Microplastics, <5 mm in length (Arthur et al., 2009), account for a large proportion of marine plastic debris and due to their small size, they are bioavailable to a wide variety of marine biota. Interactions with microplastic debris have been well documented in a number of species such as mussels, fish, marine mammals, and seabirds, however, interactions with other marine species remain poorly understood.

Ongoing research at Plymouth University is investigating the interactions between microplastics and the invasive Australasian barnacle *Austrominius modestus*. A preliminary study demonstrated that *A. modestus* ingested both polyethylene and polystyrene microplastics of four size categories: 100, 150, 200, and 250 μm. Both size and polymer type were key determinants in whether microplastics were ingested as smaller polyethylene microplastics were more readily ingested. Interestingly, rejected microplastics also appeared to become less buoyant in the experiment tanks by forming aggregates with a mucus-type substance. To further explore these aggregates, interactions will be studied under three treatments: (1) microplastics, barnacles, and sediment, (2) microplastics and barnacles, and (3) microplastics alone. Buoyancy and sinking rates of individual microplastics and aggregates will then be examined using PTV to determine if their sinking velocity has been altered.

Recently, bivalves have been highlighted as the most efficient feeders of microplastics (Setälä et al., 2016) suggesting that other filter-feeding organisms may also be ingesting high quantities of microplastics. Barnacles play a vital role in providing primary production, harvested from phytoplankton prey, to higher trophic levels. Therefore, any negative impacts associated with microplastic interactions could severely impact the functioning of important coastal ecosystems.

MICRO 2016. Fate and Impact of Microplastics in Marine Ecosystems.

Additionally, if the sinking velocity of microplastics is altered following interactions with *A. modestus*, this could present another route of microplastic contamination to deeper waters.

REFERENCES

Arthur, C., Baker, J., Bamford, H., 2009. Proceedings of the International Workshop on the Occurrence, Effects and Fate of Microplastic Marine Debris. University of Washington, Tacoma, WA, USA, 9th–11th September 2008. NOAA Technical Memorandum NOS-OR&R30.

Setälä, O., Norkko, J., Lehtiniemi, M., 2016. Feeding type affects microplastic ingestion in a coastal invertebrate community. Mar. Pollut. Bullet. 102 (1), 95–101.

Macro- and Microplastic in Seafloor Habitats Around Mallorca

C. Alomar, B. Guijarro and S. Deudero
Instituto Español de Oceanografía, Palma de Mallorca, Spain

Macro- and microplastics were assessed respectively in offshore and coastal habitats in eastern and western Mallorca. Macroplastics were sampled during annual scientific bottom trawl surveys for 15 years in both areas between 50 and 800 m using the experimental bottom trawl gear GOC-73. Contrary, microplastics were sampled in coastal sediments between 8 and 10 m by scuba divers. This study aims at giving an overview on the plastic fraction of marine debris from the micro to the macro scale at the Mediterranean Sea which is highly affected by marine litter (Deudero and Alomar, 2015).

For macroplastics, analyses revealed mean higher values $(6.11 \pm 1.30 \text{ kg/km}^2)$ in the western than in the eastern of Mallorca $(2.59 \pm 0.42 \text{ kg/km}^2)$. Contrary to this, microplastics in the eastern coastal areas showed highest concentrations, up to 0.90 ± 0.10 MPs/g, than in the western part of the Island. The eastern sampling area compromises of a Marine Protected Area (MPA) with restriction or regulation of boating activities in the coastal zone and results would be suggesting the transfer of macro- and microplastics from source areas to endpoint areas. In addition, a high proportion of microplastic filaments were found close to populated areas (western Mallorca) whereas fragment type microplastics were more common in the coastal eastern part (MPA).

This study provides with reference values for different seafloor habitats and sized plastic debris (macro and microscopic scale). Although sampling units are not standardized enabling a direct comparison between plastic concentrations between coastal and offshore seafloor habitats, it is worth noting that highest macro- and microplastic concentrations are not found in the same region. These results at different spatial levels also remark the need of sampling at multiple geographical scales in order to improve the approach to study macro- and microplastics. The data presented here is relevant and important for setting a baseline in this field of investigation which has been recognized as a problem by the United Nations which needs national, regional, and global actions.

REFERENCE

Deudero, S., Alomar, C., 2015. Mediterranean marine biodiversity under threat: reviewing influence of marine litter on species. Mar. Pollut. Bull. 98 (1–2), 58–68, 15 September 2015. <http://dx.doi.org/10.1016/j.marpolbul.2015.07.012>.

Evaluation of Microplastics in Jurujuba Cove, Niterói, RJ, Brazil

R.O. Castro[1] and F.V. de Araujo[1,2]
[1]Universidade Federal Fluminense, Niterói, Brazil [2]Universidade do Estado do Rio de Janeiro, Rio de Janeiro, Brazil

Microplastics can absorb hydrophobic compounds highly toxic and due their unbiodegradable characteristic persist in the water column and are deposited on the ocean floor, where can be ingested or filtered by organisms, entering in food chain. Jurujuba Cove, Niterói, RJ, has a large mussel farming area that serves thousand of consumers in their surroundings. To evaluate the presence of microplastics in these waters, two collections were made (in rainy (RS) and dry (DS) season). Microplastics collected through trawling, using a plankton net, were further filtered through sieves, quantified, and separated according to their size, shape, and color. Some of the fragments were chemically analyzed through the FTIR-ATR technique for chemical identification. From the 1772 fragments obtained ($16.4 \, m^{-3}$), the most abundantly size found was less than one millimeter (53% in RS and 51% in DS), typical of small plankton. Regarding to color, directly related to the adsorption of contaminants in the environment, most of the fragments collected were colorful (48% and 75% in RS and DS respectively), with the blue ones the most observed on both seasons. The plastics categorized as rigid were the most abundant, representing 57% of the samples in RS and 49% in DS. The microplastics leaf-shaped represented 22% and 24% and those classified as fibers 21% and 27% in RS and DS respectively. From the total number of leavings described, 22% presented a high degradation index. The main polymers identified were polyethylene (72%) and polypropylene (26%). The results obtained in both seasons were very similar showing that the great concentration and diversity of microplastics found were probably due to a high and constant load of effluent that this area receives and to the mussel farming activity. Understanding the amount and types of microplastics is necessary to find the sources and establish mitigating and neutralizing measures.

Presence, Distribution, and Characterization of Microplastics in Commercial Organisms From Adriatic Sea

C.G. Avio, S. Gorbi, M. Berlino and F. Regoli
Universita Politecnica delle Marche, Ancona, Italy

Microplastics (MPs) are widely diffused in the marine environment and their uptake by marine organisms is being demonstrated in a growing number of laboratory and field studies. Although several organisms can ingest MPs with potentially adverse effects, a clear picture on their distribution in wild organisms and trophic webs is still lacking.

In this work, the presence and characterization of MPs was assessed in several commercial species collected from two sites of the Central Adriatic Sea. A recently validated extraction protocol was applied to extract MPs from gastrointestinal tracts of fish and soft tissues of invertebrates. Extracted particles were characterized in terms of size, shape, and polymer typology trough microscopy and FT-IR analyses. The results indicated the MPs presence in the 45% of analyzed specimens; particles were mainly represented by fragments, pellets, and lines, while polyethylene, polystyrene, and polyamide were the dominant polymers. All the analyzed species showed a variable presence of plastic particles in their tissues. No marked differences were highlighted in the number of ingested items per specimens when comparing organisms from the two investigated sites within the same area. Invertebrates typically exhibited a lower frequency of MPs in soft tissues in respect to the stomach of fish but with a higher potential of particle transfer to human consumers.

In conclusion, this study provides new insights on the presence, distribution, and typology of MPs in commercial organisms, representing an important baseline assessment on the level of this kind of contamination in the *Adriatic* biota.

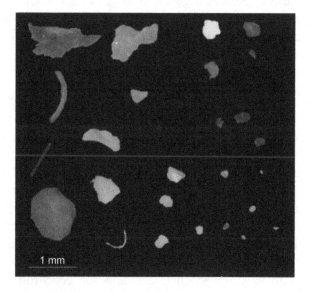

Picture of some extracted microplastics from Adriatic *species.*

The Origin and Fate of Microplastics in Saltmarshes

H. Ball, J. Kirby, E. Whitfield and K. Kiriakoulakis
Moores University, Liverpool, United Kingdom

Microplastics are rapidly becoming a global concern. They are generally characterized as plastic particles ≤ 1 mm in size (Van Cauwenberghe et al., 2013; Browne et al., 2010), and depending on the origin, can be split into primary and secondary microplastic. They are known to contaminate a number of marine environments, but they have received less attention in coastal systems. The aim of this study is to investigate the origin and fate of microplastics in saltmarsh sediments, and assess the potential for using them as a geochronological tool. Preliminary results show that microplastic pellets were present in almost all subsamples from two sediment cores taken from Parkgate saltmarsh in Cheshire, UK. There was a significant difference ($p = 0.042$; t-test) in the amount of microplastics between the two cores.

There were higher amounts of microplastics in the middle area of the marsh compared to the upper marsh area, which is expected due to more frequent inundation of water in this area. Peaks of 137 Cs, due to discharge from Sellafield nuclear reprocessing plant during the 1970s (Rahman and Plater, 2014), corresponded with peaks in microplastics. This may suggest that microplastics could be used as geochronological tools, but a more thorough and spatially representative approach is necessary to confirm this. Future work will focus on collecting further cores from radionuclide-dated saltmarshes in NW England in an attempt to evaluate the relationship of microplastic abundance with existing geochemical data for geochronological use. In addition, microplastics from these areas will be correlated with geochemical and sedimentological parameters in order to pinpoint the processes that may be related to their accumulation in the sediments.

REFERENCES

Browne, M., Galloway, T., Thompson, R., 2010. Spatial patterns of plastic debris along estuarine shorelines. Environ. Sci. Technol. 44 (9), 3404–3409.

Rahman, R., Plater, A., 2014. Particle-size evidence of estuary evolution: a rapid and diagnostic tool for determining the nature of recent saltmarsh accretion. Geomorphology 213, 139–152.

Van Cauwenberghe, L., Vanreusel, A., Mees, J., Janssen, C., 2013. Microplastic pollution in deep-sea sediments. Environ. Pollut. 182, 495–499.

Effects of Microplastics and Mercury, Alone and in Mixture, on the European Seabass (*Dicentrarchus labrax*): Swimming Performance and Subindividual Biomarkers

L.G.A. Barboza[1,2], L.R. Vieira[1,2] and L. Guilhermino[1,2]
[1]University of Porto, Porto, Portugal [2]CIIMAR/CIIMAR-LA, Interdisciplinary Centre of Marine and Environmental Research, Porto, Portugal

Considering the relevance of microplastics and mercury as global pollutants of high concern, and the lack of knowledge on their combined effects on marine fish, the main objective of the present study was to investigate the effects of these stressors, alone and in mixture, on the swimming performance and subindividual biomarkers of the European seabass (*Dicentrarchus labrax* L.), a species used for human consumption. After acclimation to laboratory conditions, 81 seabass juveniles were randomly distributed per 9 treatments (9 fish per treatment): control (filtered seawater), two concentrations of microplastics (low and high) alone, two concentrations of mercury (low and high) alone, and all the mixtures in a full factorial experimental design. Fish were exposed individually in glass beakers, with continuous aeration, and no food was provided during the 96 h. Test media was renewed at each 24 h. At the end of the exposure period, the post-exposure swimming performance of fish was assessed by determing the swimming velocity and the swimming resistance, using a device previously developed by the team. Furthermore, the following biomarkers were determined: lipid peroxidation (LPO) and the activity of the enzymes acetylcholinesterase (AChE), lactate dehydrogenase, isocitrate dehydrogenase, glutathione *S*-transferases, catalase, superoxide dismutase, and glutathione peroxidase. The concentrations of microplastics were determined in test media during the bioassay, and those of mercury in test media and in the fish body. Among other toxic effects, mercury and microplastics induced alterations in the swimming performance, decreased AChE activity and increased LPO levels. Therefore, the findings of the present study highlight the importance of further investigating the combined effects of microplastics and other common environmental contaminants in marine species, especially those used for human consumption.

DNA Damage Evaluation of Polyethylene Microspheres in *Daphnia magna*

A.A. Berber, M. Yurtsever, N. Berber and H. Aksoy
Sakarya University, Sakarya, Turkey

Polyethylene is the most produced plastic in the world, with which everyone daily comes into contact. It has wide range of applications, such as, packaging, medical, electronics, automotive, cosmetics, and more. There are many test methods used to detect potential genotoxicity of different chemicals. Among these chromosomal aberrations, both in vivo and in vitro, and single-cell gel electrophoresis (SCGE) are very sensitive methods. Hence, in the present study, we planned to determine the genotoxic potential of polyethylene microsphere (10−20 µm diameter) in *Daphnia magna* with SCGE. In order to increase knowledge on the genotoxic activity of this microplastic, three comet parameters tail length, tail DNA%, and tail moment were evaluated.

LITTERBASE: An Online Portal for Marine Litter and Microplastics and Their Implications for Marine Life

M. Bergmann, M.B. Tekman and L. Gutow
Alfred Wegener Institute, Helmholtz Centre for Polar and Marine Research, Helgoland, Germany

For many years, the pollution of the oceans with marine litter received only little attention from the public although the global plastic production has grown steadily. However, since the "discovery" of the oceanic garbage patches and microplastics the littering of the oceans has become a hot topic, which is reflected in strong recent increases in the number of publications. Despite growing research efforts many questions remain unanswered and the new wealth of information does not readily transpire to the general public, which is left unsettled. For example, it is still unclear what the overall extent of ocean pollution is, or how the enormous amounts of oceanic plastics affect marine life and ecosystems.

To overcome this uncertainty and make best use of the existing knowledge, we currently develop an online portal for marine litter and microplastic pollution named LITTERBASE. As of early 2017, LITTERBASE will provide access to the current state of understanding of marine litter and microplastics to the general public and stakeholders. Published records of marine litter and microplastics and their impact on marine life will be compiled in a database. The regularly updated information will be displayed in distribution maps and other graphs in an interactive online portal. In the long run, data from citizen scientists may also be integrated into these infographs.

The database behind the portal will further allow for scientific meta-analyses to assess the distribution of litter in our oceans and to address burning questions such as:

- "Where is all the Plastic?" (Thompson et al., 2004). Well-ground global estimates of plastic entering the oceans are currently 10–1000 times higher than results based on empirical evidence suggesting the presence of hidden/unaccounted sinks (e.g., deep seafloor, sea ice, marine life)
- Are there global patterns in the distribution of litter and microplastics?
- How is litter/microplastic partitioning in different ecosystem compartments?
- Which groups of organisms are particularly vulnerable to the impacts of marine litter and microplastics?

LITTERBASE needs support:

- Input from authors of newly published papers (e.g., pdfs) to keep LITTERBASE as complete and up-to-date as possible litterbase@awi.de
- Georeferenced photographs (from citizen scientists) documenting marine litter or showing interactions between litter and marine life to increase our knowledge base, especially from remote areas or habitats

REFERENCE

Thompson, R.C., Olsen, Y., Mitchell, R.P., Davis, A., Rowland, S.J., John, A.W.G., et al., 2004. Lost at sea: where is all the plastic? Science 304 (5672), 838.

PLASTOX: Direct and Indirect Ecotoxicological Impacts of Microplastics on Marine Organisms

A. Booth[1], K. Sakaguchi-Söder[2], P. Sobral[3], L. Airoldi[4], R. Sempere[5], J.A. Van Franeker[6], K. Magnusson[7], T. Doyle[8], I. Salaverria[9], C. Van Colen[10], D. Herzke[11], A. Orbea[12], G.W. Gabrielsen[13], H. Nies[14] and T. Galloway[15]

[1]SINTEF Materials and Chemistry, Trondheim, Norway [2]Technical University of Darmstadt, Darmstadt, Germany [3]NOVA.ID FCT, Caparica, Portugal [4]University of Bologna, Ravenna, Italy [5]Aix-Marseille Université, Marseille, France [6]IMARES Wageningen UR, Den Helder, Netherlands [7]IVL Swedish Environmental Research Institute, Göteborg, Sweden [8]National University of Ireland Galway, Galway, Ireland [9]Norwegian University of Science and Technology, Trondheim, Norway [10]Ghent University, Gent, Belgium [11]Norwegian Institute for Air Research, Tromsø, Norway [12]Universidad del País Vasco/Euskal Herriko Unibertsitatea, Leioa, Spain [13]Norwegian Polar Institute, Tromsø, Norway [14]Federal Maritime and Hydrographic Agency, Hamburg, Germany [15]University of Exeter, Exeter, United Kingdom

PLASTOX is one of four consortia under the JPI Oceans Pilot Action "Ecological Aspects of Microplastics" and consists of 15 partners from 11 Member States. The project will investigate the ingestion, food web transfer, and ecotoxicological impact of microplastics (MPs), together with persistent organic pollutants (POPs), metals and plastic additive chemicals, on key European marine species and ecosystems. It will also study the temporal dynamics of MP colonization by microbial communities and the influence of microbial biofilms on ingestion rates and POP toxicity. The influence of MP physicochemical properties (size, shape, surface area, and composition) on these processes will be evaluated. To study ecological effects of MPs, laboratory tests and mesocosm studies will be combined with field-based observations and manipulative field experiments at stations representing a wide range of European marine environments (Mediterranean, Adriatic, North, and Baltic Seas, and the Atlantic). To bridge the current gap between laboratory assessment using commercially available feedstock MPs and the additive/pollutant-loaded MPs that dominate the marine environment, macrosized plastic litter collected from the marine environment will be milled into MPs.

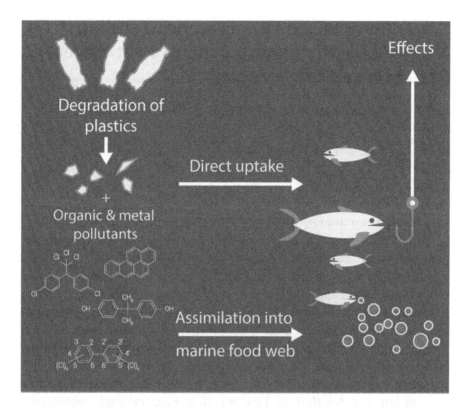

Adsorption and desorption behavior of organic and inorganic pollutants to MPs will be investigated using a range of common POP and metal contaminants, identifying which physicochemical properties are most influential. Uptake through ingestion and other routes will be investigated, and attempts made to quantify MP accumulation in marine organism tissues using state of the art analytical approaches. Acute and sublethal ecotoxicological effects of MPs will be assessed on marine organisms from phyto- and zooplankton to (shell)fish and seabirds, representative of the full range of economically important marine living resources in the EU.PLASTOX will culminate in a series of experiments bringing together the knowledge generated about MPs and POPs/metals to study the combined fate and effects of these marine contaminants in food web studies.

Community-Based Observatories Tackling MICROPLASTIC: Spanish Schools Pilot Project Based on seawatchers.org

E. Broglio[1,2], J. Baztan[1,3], O. Planells[1], P. Baztan[1,4], M. Vicioso[1] and C. Puig[5]

[1]Marine Sciences For Society [2]Institute of Marine Sciences-CSIC, Barcelona, Spain [3]Université de Versailles SQY, Versailles, France [4]Universitat Autònoma de Barcelona (UAB), Barcelona, Spain [5]Consorci El Far, Barcelona, Spain

Research data about the presence of microplastic in the environment are usually from discrete areas and hardly shared with the general public. Because the impacts of microplastic on coasts, marine ecosystems, and food webs are complex and not yet fully understood, it is difficult to convey the importance of the issue to society more broadly, despite the media's increasing interest in it. In this sense, citizen science can be a powerful tool to involve citizens in data collection and raise cultural awareness about microplastic. The website www.seawatchers.org is a citizen science platform that promotes data collection, displays georeferenced photos, and facilitates a dialog between volunteers and researchers on different research topics. In 2014 the Plastic Zero project was integrated within the platform to engage national and international groups of volunteers and allow them to report occurrences of micro- and macroplastics in their areas. In 18 months we collected approximately 300 photos of micro- and macroplastic from citizens and environmental organizations all over the world. After this period of diagnosis, we identified the need to collect time series data and from there arose the aim of building "microplastic observatories" based on citizen participation to monitor the presence of microplastic throughout the year in given locations.

In December 2015 we started a pilot project with three schools in the Barcelona area. This pilot served to (1) implement and improve a sampling protocol to be followed by citizen scientists; (2) identify the best places for sustainable sampling stations, and (3) integrate new partners in the project for the next steps. The results of this 3-month pilot study indicate that schools are a very good target for this purpose, and this collaboration has great potential to provide good data for scientific projects as well as to increase awareness and action for young people.

Tackling Marine Litter: Awareness and Outreach in the Azores

R. Carriço[1], J.P.G.L. Frias[1], Y. Rodriguez[2], N. Rios[2], M.J. Cruz[2], C. Dâmaso[2] and C.K. Pham[1]

[1]Universidade dos Açores, Horta, Portugal [2]OMA – Observatório do Mar dos Açores, Horta, Portugal

Marine litter is a global problem, affecting several ecosystems and a countless number of species, with direct socioeconomic and environmental impacts on tourism and fisheries.

Environmental awareness and public outreach actions are key behavior change measures, as they target practically any sector of society, and contribute to minimize impacts, through the use of media (general public) or targeted education campaigns to a particular audience (e.g., school) on a specific issue.

In the Azores archipelago (north-eastern Atlantic), numerous educational and awareness activities focused on marine litter have been carried out by the Observatório do Mar dos Açores (OMA), since 2001, mainly on Faial island (~15,000 inhabitants). Since 2015, OMA has been collaborating with a team of researchers from an IMAR-DOP project (Azorlit— Establishing a Baseline on Marine Litter in the Azores), combining the technical know-how of OMA with the scientific knowledge of IMAR.

So far, the number of activities developed by both organizations together has reached over a 1000 people. One particular campaign that has been particularly successful is an underwater cleanup action (2001–15), carried every year in Horta harbor. In addition, many actions were undertaken at local schools and during summer camps, for which specific games related to the issues were created.

The high number of people reached in the island makes these activities successful with strong engagement of local people, particularly children, who act as vectors of change and cause a wider impact in society.

Dissemination of environmental education actions and related information are important when easily communicated to local people in a simple and understandable way. This partnership aims to develop educational marine litter programs, based on scientific data collected in the archipelago and translated to local people through a science communication process.

Prevalence of Microplastics in the Marine Waters of Qatar's Exclusive Economic Zone (EEZ)

A. Castillo, I. Al-Maslamani and J.P. Obbard

Qatar University, Doha, Qatar

Although the potential threat of microplastics on the marine environment, and associated ecosystems, is now well recognized, there is no baseline data available for the Arabian Gulf. The Environmental Science Center of Qatar University has now documented the first evidence for the prevalence of microplastics within marine waters of the Arabian Gulf, specifically in the coastal marine waters of Qatar. Qatar has an arid climate and is situated midway along the western coast of the semienclosed Arabian Gulf. Qatar's coastline is particularly susceptible to marine debris due to the country's rapid urbanization and economic development.

Surface marine water samples were collected from 12 marine stations during May 2015 research cruise onboard RV Janan. An optimized and validated protocol was developed for the extraction of microplastics from plankton-rich seawater samples without loss of microplastic debris present. Following extraction, microplastics were characterized using Attenuated Total Reflectance-Fourier Transform Infrared (ATR-FTIR) spectroscopy. Microplastics were identified in ten out of twelve seawater samples, with an average concentration of 0.71 particles/m^3 (range 0–3 particles/m^3). Polypropylene, low density polyethylene, polyethylene, polystyrene, polyamide, polymethyl methacrylate, cellophane, and acrylonitrile butadiene styrene microplastic particles were identified, and ranged in size from 125 μm to 1.82 mm for granular particles, and 150 μm to 15.98 mm for fibrous shaped microplastics.

Microplastics showed evidence of oxidation, where exposure to the high salinity of Qatar's marine environment, as well as elevated levels of ultraviolet radiation, create an aggressive environment for the degradation of plastic debris into microplastics. The microplastics are evident in areas where nearby anthropogenic activities, including oil-rig installations and shipping operations are present.

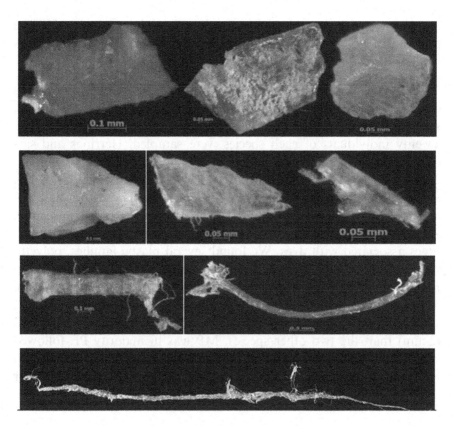

Microplastics isolated from the marine waters of the Qatar's EEZ.

Microplastics Contamination in Three Planktivorous and Commercial Fish Species

F. Collard, K. Das, B. Gilbert, G. Eppe and E. Parmentier
[1]University of Liege, Liège, Belgium

In 2014, 311 million tons of plastics were produced worldwide while it is estimated that 10% ended up in the oceans. Plastics adsorb many pollutants, cause entanglements in many vertebrates and provoke blockage of the digestive tract of marine mammals, birds, and turtles. Plastics also fragment in smaller parts, forming microplastics (<5 mm). These microplastics have the same size than planktonic organisms and can thus be ingested by filter-feeders and planktivorous organisms including fish. Although planktivorous Clupeiforme species are the most fished species in the world (17 million tons in 2013), the possible impact of microplastic ingestion on this group has received little attention.

The aim of this study was to quantify and to characterize the ingested anthropogenic particles (AP) in herrings (*Clupea harengus*), sardines (*Sardina pilchardus*), and anchovies (*Engraulis encrasicolus*). Twenty individuals of each species were sampled. Herrings and sardines were caught in the Channel and in the North Sea in January 2013 and in January 2014. Anchovies were sampled in July 2013. The stomach contents were digested by sodium hypochlorite to isolate AP and then they were analyzed by Raman spectroscopy and measured. AP were constituted of microplastics (MP) and cellulose fibers, sometimes associated with dyes. We found several plastic families: polyethylene (PE), polypropylene (PP), polystyrene, polyacrylonitrile, polyethylene terephthalate, polyamide, and poly(butyl methacrylate). MP occurred in 35%, 50%, and 40% of stomach contents of anchovies, sardines, and herrings, respectively. Anchovies ingested bigger AP (2 mm) than the two other species. MP was in majority PE followed by PP. Our study shows that highly commercial species are exposed to MP. Further research is needed as MP are transport medium for organic pollutants and their fate once in the organism is unknown.

Spatial Variation of Microplastic Ingestion in *Boops boops* in the Western Mediterranean Sea

M. Compa, C. Alomar, A. Ventero, M. Iglesias and S. Deudero
Instituto Español de Oceanografía, Madrid, Spain

Microplastics are of increasing concern due to their widespread presence in global marine environments. Whether they are entering the systems as fibers or pellets or whether they are broken down and fragmented from larger marine plastics, evidence of accidental ingestion of microplastics by marine fauna, especially fish, is increasing. This study aims to describe the presence of microplastics in the gastrointestinal tract of the demersal fish species *Boops boops* from various locations in the Western Mediterranean Sea. Samples collected from the Balearic Islands were collected from bottom trawl

Survey	Location	Total MPs	als with MPs	Number of Individuals	Gestion MPs	Length (mm) (Mean ± SD)	Weight (g) (Mean ± SD)
MEDIAS	Almeria	4	3	13	1.3 ± 0.58	237.7 ± 6.7	106.9 ±
	Garrucha	2	2	11	1 ±	259.5 ± 10.6	151 ±
	Barcelona	2	2	15	1 ±	152.5 ± 53.0	37.2 ±
	Motril	11	6	10	1.8 ± 0.8	223.2 ± 9.8	90.1 ±
Commer-fishing	Ibiza Llevant	27	74	93	3.7 ± 3.9	193.4 ± 22.5	77.2 ±
	Ibiza Ponent	12	51	67	2.5 ± 1.7	197.5 ± 16.3	81.0 ±
	Ratjada	23	51	100	4.7 ± 4.1	190.4 ± 17.7	67.8 ±
	Palma	93	19	77	4.9 ± 1.9	172.2 ± 25.4	50.6 ±

Table 1 Summary of Microplastic Ingestion for *Boops boops*

commercial fishing vessels and samples from the south-eastern coast of Spain were collected during the oceanographic acoustic survey MEDIAS.

Of the fish that had ingested plastic, both locations in Ibiza had the highest percentage of ingestion (Ibiza Llevant 19.5%; Ibiza Ponent 76.1%) and the Barcelona site from the MEDIAS survey had the lowest percent of ingestion (13%; Table 1). Although ingested rates varied by location, ingested microplastics were found at all of the locations, indicating MPs are common in the food web of *Boops boops* in small amounts regardless of location.

Source, Transfer, and Fate of Microplastics in the Northwestern Mediterranean Sea: A Holistic Approach

C. Mel, K. Philippe and H. Serge
Université de Perpignan, Perpignan, France

Numerous, recent studies have shown the ubiquitous presence of microplastic (MPs) fragments at the seawater surface. However, a number of questions remain concerning the sources of these fragments and their fate once introduced into the marine environment. We propose a holistic sampling strategy designed to cover the entire cycle of MPs within the marine environment—from the source to the final sink—and aiming to provide some answers to the following questions:

- What are the respective contributions of rivers and atmosphere to the MP input into the sea?
- What are the main controlling factors of MP input into the sea: river size and outflow, type of land coverage (urban, farming, industries), climatic events,...?
- What is the fate and behavior of MPs once introduced in the sea: drift towards the open ocean and/or the beach, settling to the seafloor, ingestion by biota?
- Where are the final sinks and what is the proportion of MPs ultimately buried there?

Our experimental work focuses on a French coastal area: the Gulf of Lions (NW Mediterranean Sea). Regarding MP input, we consider the Rhône River, major inflow to the area as well as the Têt River, a small typical Mediterranean river, both being sampled on a month basis with a Manta trawl ($> 300\ \mu m$). We also determine atmospheric MP deposition on the roof of our laboratory, sampled after rain events. To identify transfer and sinks we sample river sediments, sea surface water around the Rhône and Têt river mouths (monthly basis), two beaches, either impacted or not by the proximity of a river (monthly), settling particles collected by sediment traps (archives), coastal sediments (seasonal), and deep marine sediment (archives). The study started in October 2015 and will last until December 2016.

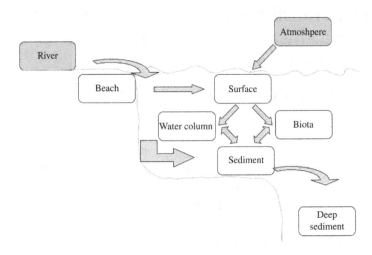

Schema describing the holistic approach. Turquoise squares: potential sources of MPs; transparent squares: potential sinks of MPs; arrows: transfer of MPs.

Microplastics in the Final Ocean Frontier

W. Courtene-Jones[1], B. Quinn[2], S.F. Gary[1] and B.E. Narayanaswamy[1]

[1]Scottish Association for Marine Science, Oban, Scotland [2]University of the West of Scotland, Paisley, Scotland

Microplastics, plastics <5 mm in diameter, are found throughout the marine environment and pose one of the greatest threats to the world oceans health. Numerous species are documented to ingest microplastics suffering negative impacts as a consequence. While microplastic research has developed over the last decade, significant knowledge gaps still exist. Current estimates suggest a large proportion of microplastics are unaccounted for in the ocean, indicating the deep sea may be a reservoir for this missing fraction. The deep sea is one of the richest ecosystems in the world, yet due to its inaccessibility limited research has been conducted here, however microplastics have been identified in deep-sea sediments. I aim to provide the first assessment of the presence of microplastics in deep-sea benthic invertebrates. A rapid extraction technique, based on enzymatic digestion, has been developed to examine microplastics in faunal tissues. Furthermore, an evaluation of specimen preservation techniques (freezing and formaldehyde/ethanol) for viability and application in microplastic research has been carried out. From these enquires I will elucidate when microplastics began to appear in the deep sea, using a historical specimen time-series dating back to the 1970s. Finally, little is known about the fate of microplastics in the marine environment or their final resting place. A series of computational models will examine potential sources of microplastics in the Northeast Atlantic deep sea, considering vertical transport routes, sinking rates, and residence times of particles. Here I present preliminary work and ongoing multidisciplinary research to further our understanding of microplastics in the deep sea.

Microplastic Abundance and Distribution in the Intertidal and Subtidal Marine Environment Around a Major Urban Park in Vancouver, Canada

A. Díaz[1,2], E. Gies[1], E. Crichton[1], M. Noel[1], K. Le Souef[3], I. Riba[2] and P.S. Ross[1]

[1]Vancouver Aquarium Marine Science Centre, Vancouver, BC, Canada [2]Universidad de Cádiz, Cádiz, Spain [3]Vancouver Aquarium, Vancouver, BC, Canada

For further information on this study, please contact the authors.

Microplastic Ingestion by Decapod Larvae

K. Reilly[1], E. Fileman[1], A.W. McNeal[1], P. Lindeque[1] and M. Cole[1,2]

[1]Plymouth Marine Laboratory, Plymouth, United Kingdom [2]University of Exeter, Exeter, United Kingdom

Microplastics are a widespread and increasingly acknowledged problem in the marine environment and are reported to have detrimental impacts on the feeding, health, and survival of a range of marine biota. Decapod larvae are the juvenile form of a range of commercially important crustacean species and also play a fundamental ecological role in the ecosystem. We investigated the uptake and impact of microplastics on decapod larvae commonly found in the western English Channel. First, we demonstrated the capacity for a range of decapod larvae to ingest microplastics; second, we conducted a feeding experiment to ascertain any significant impact on decapod larval feeding when exposed to microplastics and third we used enzymatic digestion methods to reveal the presence of microplastics in field samples of a range of decapod larvae from the western English Channel. The results of these investigations will be discussed.

May Polystyrene Microparticles Affect Mortality and Swimming Behaviour of Marine Planktonic Invertebrates?

C. Gambardella[1], S. Morgana[1], M. Bramini[2], F. Misurale[1], S. Ferrando[3], V. Piazza[1], F. Garaventa[1] and M. Faimali[1]

[1]Institute of Marine Science (ISMAR), Genova, Italy [2]IIT, Italian Institute of Technology, Genova, Italy [3]University of Genova, Genova, Italy

Plastic debris is accumulating in the environment, fragmenting into small pieces known as microplastics (MPs). This kind of marine litter may be easily ingested by both vertebrates and invertebrates, accumulating through the food chain. The aim of this study was to verify whether 0.1 μm polystyrene particles may affect acute and behavioral responses in marine planktonic invertebrates. Two end-points, mortality, and swimming speed alteration, were investigated in the larval stages of the barnacle *Amphibalanus amphitrite* and of the brine shrimp *Artemia franciscana* and in the rotifer *Brachionus plicatilis* by exposing the organisms to a wide range of MP concentrations (from 0.001 up to 100 mg/L) for 24 and 48 h. In addition, MP organism uptake and the state of MP aggregation in the exposure medium were investigated by means of microscopical analysis and Dynamic Light Scattering (DLS) during both exposure times.

The results show that MPs tended to aggregate in seawater and to accumulate in all marine invertebrates after 24 and 48 h, without inducing a toxic effect on mortality and swimming alteration in any species after 24 h ($LC_{50}/EC_{50} > 100$ mg/L). Acute and behavioral end-points were only affected in barnacle nauplii and brine shrimp larvae exposed for 48 h at MP concentrations much higher than those estimated for the marine water (from 1 ppm onwards).

In conclusion, 0.1 μm polystyrene particles did not cause any ecotoxicological effects in marine planktonic invertebrates at environmental concentrations; however, the ingestion observed in all organisms need to be further investigated analyzing their possible effect at cellular and/or molecular level.

Improvement of Microplastic Extraction Method in Organic Material Rich Samples

P. Garrido, A. Štindlová, A. Herrera and M. Gómez
Universidad de Las Palmas de Gran Canaria, Las Palmas de Gran Canaria, Spain

Microplastics are small plastic particles that, because of their size, are available for many marine organisms. Their consumption can be hazardous and have many chemical, mechanical, and biological effects. Nowadays, the scientific community is focused on determining the microplastics abundance and distribution to gain a better understanding of the magnitude of this global problem. Prior to that, it is essential to harmonize sampling, extraction, and quantification methodologies in order to get reliable and reproducible data of microplastic contamination in marine biota. As many microplastic samples include biological material that could mask the presence of the plastic particles, it must be eliminated before proceeding with the analysis. Although there are several methodologies described for the digestion of zooplankton within microplastic samples, methodologies for the digestion of vegetal and algal material are not so common. In this study, we focus on the digestion of algae and vegetal material, and compare the efficacy of two methodologies, described for the microplastics extraction in organic material rich samples, in the digestion of algae and vegetal material from sediment samples. Preliminary results show that HCl and NaOH are not effective in the digestion of this particular type of organic material, and that other methodologies should be studied.

Defining the Baselines and Standards for Microplastics Analyses in European Waters (JPI-O BASEMAN)

G. Gerdts[1], K. Thomas[2], D. Herzke[3], M. Haeckel[4], B. Scholz-Böttcher[5], C. Laforsch[6], F. Lagarde[7], A.M. Mahon[8], M.L. Pedrotti[9], G.A. de Lucia[10], P. Sobral[11], J. Gago[12], S.M. Lorenzo[13], F. Noren[14], M. Hassellöv[15], T. Kögel[16], V. Tirelli[17], M. Caetano[18], A. Collignon[19], I. Lips[20], O. Mallow[21], O. Seatala[22], K. Goede[23] and P. Licandro[24]

[1]Alfred Wegener Institute Helmholtz Centre for Polar and Marine Research (AWI), Helgoland, Germany [2]Norwegian Institute for Water Research (NIVA), Oslo, Norway [3]Norwegian Institute of Air

Research (NILU), Tromso, Norway [4]GEOMAR Helmholtz-Zentrum für Ozeanforschung, Kiel, Germany [5]University of Oldenburg, Institute for Chemistry and Biology of the Marine Environment (ICBM), Oldenburg, Germany [6]University of Bayreuth, Bayreuth, Germany [7]University of Maine, Le Mans, France [8]Institute of Technology, Galway, Ireland [9]CNRS-LOV, Villefranche sur Mer, France [10]CNR-IAMC, Oristano, Italy [11]NOVA.ID FCT, Caparica, Portugal [12]Instituto Español de Oceanografía (IEO), Vigo, Spain [13]Universidade da Coruña (UDC)-Instituto, A Coruña, Spain [14]IVL Swedish Environmental Research Institute, Fiskebäckskil, Sweden [15]University of Gothenburg, Gothenburg, Sweden [16]The National Institute of Nutrition and Seafood Research (NIFES), Bergen, Norway [17]OGS- National Institute of Oceanography and Experimental Geophysics, Trieste, Italy [18]Instituto Português do Mar e da Atmosfera, Lisbon, Portugal [19]University of Liege, Liège, Belgium [20]Marine Systems Institute at Tallinn University of Technology, Tallinn, Estonia [21]Vienna University of Technology, Vienna, Austria [22]Finnish Environment Institute, Helsinki, Finland [23]Rap.ID Particle Systems GmbH, Berlin, Germany [24]Sir Alister Hardy Foundation for Ocean Science (SAHFOS), Plymouth, United Kingdom

Since the middle of last century rapidly increasing global production of plastics has been accompanied by an accumulation of plastic litter in the marine environment. Dispersal by currents and winds does not diminish the persistence of plastic items which degrade and become fragmented over time. Together with microsized primary plastic litter from consumer products these degraded secondary microfragments lead to an increasing amount of small plastic particles (smaller than 5 mm), so called "microplastics." The ubiquitous presence and massive accumulation of microplastics in marine habitats and the uptake of microplastics by various marine biota is now well recognized by scientists and authorities worldwide. Although awareness of the potential risks is emerging, the impact of plastic particles on aquatic ecosystems is far from understood. A fundamental issue precluding assessment of the environmental risks arising from microplastics is the lack of standard operation protocols (SOP) for microplastics sampling and detection. Consequently there is a lack of reliable data on concentrations of microplastics and the composition of polymers within the marine environment. Comparability of data on microplastics concentrations is currently hampered by a huge variety of different methods, each generating data of extremely different quality and resolution. Although microplastics are recognized as an emerging contaminant in the environment, currently neither sampling, extraction, purification nor identification approaches are standardized, making the increasing numbers of microplastics studies hardly—if at all—comparable. BASEMAN is an interdisciplinary and international collaborative research project that aims to overcome this problem. BASEMAN teams experienced scientists (from different disciplines and countries) to undertake a profound and detailed comparison and evaluation of all approaches from sampling to identification of microplastics. BASEMAN's project outcomes will equip policy makers with the tools and operational measures required to describe the abundance and distribution of micro-plastics in the environment. Such tools will permit evaluation of member state compliance with existing and future monitoring requirements. BASEMAN is one of four projects funded in the framework of the JPI-O pilot action "Ecological Aspects of Microplastics."

Effects of Long-Term Exposure With Contaminated and Clean Microplastics on *Mytilus edulis*

T. Hamm

GEOMAR, Helmholtz Center for Ocean Research in Kiel (GEOMAR), Kiel, Germany

For further information on this study, please contact the authors.

The City of Kuopio and Lake Kallavesi in the Finnish Lake District—A "Living Laboratory" for the Microplastic Pollution Research in Freshwater Lakes

S. Hartikainen, T. Bizjak, T. Gajst, J. Leskinen, P. Pasanen, A. Koistinen and J. Sorvari

University of Eastern Finland, Kuopio, Finland

The effects of microplastic pollution in freshwater basins remain generally understudied in Finland. The majority of lakes in Finland are located to the Finnish Lake District in the eastern part of Finland. The hilly landscape of the lake plateau is dominated by glacial remnants of drumlins and by long sinuous eskers. Terminal moraines trap networks of thousands of lakes separated by hills and forested countryside. There are 187,888 lakes in Finland. With an area of 4728 km², Lake Kallavesi is the tenth largest lake and is located in the region of Northern Savo in eastern Finland. It belongs to the Vuoksi main catchment area, which covers an area of 16,270 km². Vuoksi is the largest freshwater catchment area in Finland. Lake Kallavesi surrounds the City of Kuopio, which is the eight biggest city in Finland with population of 112,000. The main aim of this research is to turn the City of Kuopio into a "living laboratory" focusing on how urban waste and drainage waters and littering affect freshwater ecosystems such as Lake Kallavesi. The "Living laboratory of Kuopio" together with the state-of-the-art analytical laboratories at the Kuopio campus of the University of Eastern Finland provides opportunities to carry out on-site research of sources, fate, and effects of microplastics in Lake Kallavesi. Furthermore, the research aims to assess the effect of the stark contrasts of four seasons on the abundance and distribution of microplastic in Lake Kallavesi. For almost 5 months the Lake Kallavesi is covered with ice which could retain the otherwise floating particles on the surface. Our preliminary results of the ice samples confirm the presence of microplastic in Lake Kallavesi with a substantial amount of fibers.

Analysis and Quantification of Microplastics in the Stomachs of Common Dolphin (*Delphinus delphis*) Stranded on the Galician Coasts (NW Spain)

A. Hernandez-Gonzalez[1], C. Saavedra[1], J. Gago[1], P. Covelo[2] and M.B. Santos[1]

[1]Instituto Español de Oceanografía, Vigo, Spain [2]Coordinadora para o Estudio dos Mamíferos Mariños, CEMMA, Gondomar, Spain

In Galicia (NW Spain) the nongovernmental organization CEMMA assist stranded cetaceans since 1990. Over the last 26 years, in this area the common dolphin (*Delphinus delphis*) is the most frequent stranded cetacean and a total of more than 2700 common dolphins have been recorded so far. Stomachs of stranded cetaceans in good conservation state are routinely collected for diet analysis. Due to most marine debris (composed mainly of plastics) has been recognized as a global threat to marine life because plastic ingestion can lead to injuries derived from chemical exposure, we investigated the presence of plastic debris in the stomach content of 25 common dolphins stranded on the coast of Galicia between 2005 and 2010. To date, very few studies of this kind have been carried out in cetaceans even if studies of the digestive tract of several marine species have reported presence of marine debris. Based on the available literature of cetaceans and other biota (fish, seabirds, and turtles) we developed a protocol to determine and quantify the presence of plastic debris in cetacean stomach contents. Stomach contents were rinsed and organic material digested for visual inspection.

We found and isolated a total of 411 plastic items which were visually identified through a stereoscopic microscope. Although debris abundance varied from one stomach to another ($XX = 12 \pm 8$), microplastics were present in 100% of the stomachs of common dolphins analyzed. Most of items were small plastic fibers (96.59%) although plastic fragments (3.16%), and pellets (0.25%) were also identified into the analyzed stomachs. Although the amount and size of plastics do not seem to be enough to compromise the functioning of the digestive tract, could act as vectors of toxic pollutants causing endocrine effects, affecting the metabolism or other biological processes.

First Report of Microplastic in Bycaught Pinnipeds

G. Hernandez-Milian[1], A. Lusher[2,3], S. MacGabban[1], M. Gosch[1,4], I. O'Connor[3], M. Cronin[4] and E. Rogan[1,5]
[1]University College Cork, Cork, Ireland [2]National University of Ireland-Galway, Galway, Ireland [3]Galway-Mayo Institute of Technology, Galway, Ireland [4]Coastal & Marine Research Centre (CMRC), UCC, Cork, Ireland [5]Irish Whale and Dolphin Group, Kilrush, Ireland

Large numbers of bycaught seals have been reported in set-net gears along south and west coasts of Ireland; these seals are a good source of data to investigate the ecology and health status of one of the most common top predators in Irish waters. Between 2012 and 2015, 14 digestive tracts (stomach and intestines) from bycaught seals recovered from trammel nets targeting monkfish and rays off the south coast were analyzed. Routinely, the diet analysis and parasitic infection were performed (see other presentations), and as a novel study anthropogenic debris occurrence was also investigated. Microplastic separation and identification was carried out following previous methodology used in other marine mammal studies in Ireland. The incidence of microplastic was found to be over 80% of the seals, were fibers were the most common microplastic item found. We hypothesized two sources of microplastic ingestion: (1) trophic transfer through their prey, and (2) the ingestion of net fibers from the nets as they have been reported actively depredating fish from static gear.

The Contribution of Citizen Scientists to the Monitoring of Marine Litter

V. Hidalgo-Ruz[1] and M. Thiel[1,2]

[1]Universidad Católica del Norte, Coquimbo, Chile [2]Centro de Estudios Avanzados en Zonas Áridas (CEAZA), Coquimbo, Chile

Citizen science are scientific projects supported by volunteer participation of untrained citizens. Herein we provide an overview of the marine litter studies that have been supported by citizen scientists, comparing these studies ($n = 36$) with selected studies conducted by professional scientists ($n = 40$). Citizen science have been made mainly to determine the distribution and composition of marine litter of the intertidal zone; on regional, national, and international scales; and covering time periods ranging from less than one year up to 2 years. Professional studies have also been made mainly to determine distribution and composition of marine litter, but from the intertidal, subtidal, and pelagic zones, with a part of them focused exclusively on microplastics. They have been done over local, regional, and international scales, and covering mainly a spatial scale of less than one year. Both citizen science and professional studies on marine litter have been conducted mainly in the Northern Hemisphere, revealing a lack of information for coastal regions of the Southern Hemisphere. A main concern of marine litter science is to assure the quality of the collected data. Every study should include a data-quality process, including preparation of clear protocols, training of volunteers, in situ supervision by professional scientist, and revision of samples and data. New research topics for citizen science can be addressed in the future, for instance, with divers and sailing associations. Citizen science have been prove to be a useful approach to expand the understanding of marine litter, increasing the available information in this field.

Experimenting on Settling Velocity of Cylindrical Microplastic Particles

I.A. Isachenko, L.I. Khatmullina and I.P. Chubarenko
Atlantic Branch of P.P. Shirshov Institute of Oceanology, Russian Academy of Sciences, Kaliningrad, Russia

An assessment of the settling velocity of different classes of microplastic particles (<5 mm) is essential to propose a reasonable parametrization of their vertical transport. Settling velocity of natural sediments has been extensively studied in a wide range of Reynolds numbers by numerous researchers. Properly calibrated semiempirical models demonstrate good results in predicting the settling velocity of natural particles of different shape and roundness. Since small plastic debris may exhibit broader variability of shapes than the naturally occurring grains, it is questionable if the approaches used in sedimentology could be applicable to the microplastics.

Results of experiments on settling velocity of elongated cylinders are reported. Most of the particles were artificially made from fishing lines and nets of different diameters. Preliminary experiments revealed discrepancy between the measured values and those predicted by several semiempirical models—this needs further investigation. Additionally we made an attempt to estimate the effect of biofouling on settling using the naturally exposed material (found in coastal waters).

The research is supported by the Russian Science Foundation, project number 15-17-10020.

WEATHER-MIC—How Microplastic Weathering Changes Its Transport, Fate, and Toxicity in the Marine Environment

A. Jahnke[1], M. MacLeod[2], A. Potthoff[3], E. Toorman[4] and H.P. Arp[5]

[1]Helmholtz Centre for Environmental Research – UFZ, Leipzig, Germany [2]Stockholm University, Stockholm, Sweden [3]Fraunhofer Gesellschaft, IKTS, Dresden, Germany [4]KU Leuven, Leuven, Belgium [5]Norwegian Geotechnical Institute (NGI), Oslo, Norway

Understanding the hazards posed by microplastics in the sea requires understanding the changes they undergo as a result of environmental weathering processes, like UV exposure, biofilm growth, and physical stress. These processes influence parameters such as their brittleness, density, size, and surface charge, which can in turn affect their environmental fate as the microplastics undergo fragmentation, aggregation, and ultimately sedimentation or mineralization. As these processes occur, there are a series of tradeoffs of hazard to the marine environment. Changes that lead to fragmentation or mineralization into benign fragments or molecules reduce potential hazards; whereas changes that lead to the production of problematic size fractions (e.g., that can accumulate in gills) and release toxic chemicals increase potential hazards. Similarly, the influence on mobility is also wide-ranging, as some fragments may be soluble while others form aggregates that settle on the seabed.

The WEATHER-MIC project aims to develop novel tools for tackling the complex implications of weathering of microplastics in a holistic manner. The toolbox of analytical and (eco)toxicological methods and models includes (1) "fingerprinting" methods to track microplastic weathering, (2) mechanisms of chemical release from microplastics, (3) advanced particle imaging to investigate size distribution and morphological changes with weathering, (4) improved understanding regarding the biofilm that accumulates on microplastics and its trophic transfer, (5) hydrodynamic models to account for changes in sedimentation and transport with microplastic fragmentation-aggregation, and (6) (eco)toxicity profiles for weathered microplastics. The scientific products of WEATHER-MIC will provide a basis to integrate weathering of microplastics into risk assessments of marine plastic debris, considering both exposure and effects. The novel tools will open avenues to improved guidelines for plastic handling, safeguards that can minimize the hazards of microplastic pollution, and public education that aims to reduce the levels of plastic litter in the marine environment of Europe and globally.

Outline of the research aspects tackled within the WEATHER-MIC project.

Uptake of Textile Polyethylene Terephthalate Microplastic Fibers by Freshwater Crustacean *Daphnia magna*

A. Jemec[1], P. Horvat[2] and A. Kržan[2]

[1]University of Ljubljana, Ljubljana, Slovenia [2]National Institute of Chemistry, Ljubljana, Slovenia

Textiles are a recognized source of microplastics released into the freshwater environment via sewage discharges. However, very little information is currently available on the ingestion and effects of textile-derived microplastic fibers on freshwater organisms. In the current work we focused on the effect on the freshwater crustacean water flea *Daphnia magna* which is one of the dominant members of freshwater zooplankton and it is commonly used as a fundamental model in ecotoxicology and aquatic ecology. We investigated the ingestion of polyethylene terephthalate textile fibers (polydisperse size range, length $0.062-1.273$ mm, width $0.03-0.528$ mm, and area $0.004-0.128$ mm^2) by *D. magna* and the potential effects of MP ingestion on the molt, growth, and survival of daphnids after 48 h of exposure, and survival after subsequent 24 h recovery in microplastic free medium. We provide evidence that: *D. magna* is able to ingest large microplastic fibers above up to 300 μm; which is contrary to general belief that these organisms (size 1 mm) are able to ingest spherical plastic particles with an upper diameter limit of 50 μm. Acute exposure to microplastics may result in increased mortality of these crustaceans; but no effect on the growth and molt of daphnids was evidenced after 48 h of exposure. Microplastic particles are commonly attached onto the surface of the organisms. It remains to be investigated how chronic exposure may affect this small freshwater zooplankton species.

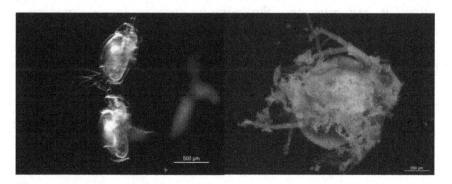

Water flea Daphnia magna *with full gut of microplastic (left) and covered with microplastic fibers (right).*

Microplastics Deposited on Two Beaches of the Gulf of Lions (NW Mediterranean Sea): Study of the Spatial Heterogeneity

K. Philippe[1], M.-V.-V. Morgan[1,2], C. Mel[1] and D. Marc[2]

[1]Université de Perpignan, Perpignan, France [2]Parc naturel marin du Golfe du Lion, Port-Vendres, France

Litter is one of the criteria used to assess the ecological state of the marine environment. It is however difficult to evaluate and compare concentrations of small particles such as microplastics (MPs), an important, ubiquitous category of marine litter, due to the plethora of sampling methodologies used to estimate MP abundance on beaches for example. To define a sampling strategy suitable for a microtidal sea (Mediterranean Sea) and for a monthly monitoring by the French Marine Protected Area Agency, we studied MP distribution on two contrasted beaches from the northwestern Mediterranean Sea (Gulf of Lions): a beach within a rocky shore far from large rivers ("Plage du Fourrat," Paulilles Bay, France) and a beach close to the mouth of a typical coastal Mediterranean river ("Plage de La Crouste," Canet, France). A high-resolution strategy was used to map the heterogeneous distribution of MPs along three distinct, parallel deposition lines (low, middle, and high) at four locations along a North–South transect. Despite the lack of tides, storms and strong winds create deposition lines of different ages across the beach. Our results show an increase of MP concentrations from the most recent deposition line (low level) to the oldest one (high level). MP concentrations also increase from north to south likely because of the local N–S alongshore drift. The choice of a deposition line thus strongly influences MP estimates. Furthermore, high-resolution sampling is time-consuming and does not easily fit a routine monthly sampling. The main challenge here will be to find the optimal sampling effort reduction while still considering the MP spatial heterogeneity on local beaches.

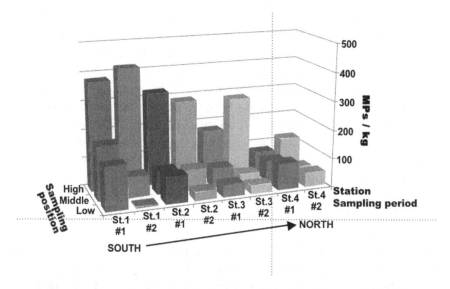

MP distribution (number per m²) along the Fourrat beach (Baie de Paulilles, Port-Vendres, NW-Mediterranean Sea, France) along three parallel deposition lines across the beach (low, middle, and high) at four North–South locations.

Effects of Microplastics on Digestive Enzymes in the Marine Isopod *Idotea emarginata*

Š. Korez, L. Gutow and R. Saborowski

Alfred Wegener Insitute Helmholtz Centre for Polar and Marine Research (AWI), Bremerhaven, Germany

Light weight, mechanical resistance, and low production cost are some of the positive features of plastic materials which gained versatile use over recent decades. Growing human population and careless usage of plastic products along with increasing global plastic production, lead to massive litter accumulations in natural environments. Due to river runoff and gradual degradation of larger plastic pieces into microsized ones microplastics enter the marine environment. Although we are aware since the early 1970s of the threat plastic is representing, there is still little information about the effects of microplastic on marine biota. In a recent study it was demonstrated that microplastic is ingested and excreted by the marine isopod *Idotea emarginata* without clogging the digestive system of the animals. In the present study we are now studying the effect of microplastics on feeding rates and physiology in this abundant invertebrate group from sub- and eulitoral habitats. Applying different feeding treatments with the fresh brown algae *Fucus vesiculosus* and dried algae embedded in agarose we studied the effect on digestive enzyme activities in the gut and midgut gland when the food was blended with microplastic. Feeding rates did not change when microplastics were added to the diet. Enzyme activities showed high scatter and inconsistent results. Esterase (C4) and lipase (C18) activities decreases in the gut and in the midgut gland when fresh algal food was enriched with microplastics. Activities of phosphoesterase exo- and endopeptidase showed no distinct changes in activity. Our results provide first evidence that microplastic may differently affect digestive enzyme activities in the gut and the midgut gland of marine isopods. Further research is needed to verify whether the observed alterations affect food digestion and nutrient assimilation.

Sinking Behavior of Microplastics

N. Kowalski, A.M. Reichardt, M. Glockzin, S. Oberbeckmann and J.J. Waniek
Leibniz Institute for Baltic Sea Research, Rostock, Germany

Plastic debris <5 mm is defined as microplastics and is, meanwhile, wide-spread in our marine ecosystems. Due to the potential to carry organic and inorganic pollutants as well as harmful microorganisms, microplastics rapidly became the focus of numerous marine pollution studies. Diverse types of polymers with different physical properties are found in water column and sediments of coastal areas. However, little is known about the behavior of microplastics, their pathways and distribution in natural aquatic systems.

Here, results from laboratory experiments are presented using sedimentation columns and high-resolution backlit imaging together with subsequent digital image analysis (Glockzin et al., 2014) in order to study sinking velocities of polymer particles in water with different salinity. We chose a range of polymers with different densities (PS, PA, PMMA, PET, POM, and PVC) which are commonly used in the plastics production and probably end up as waste in our ecosystems. The results show that the sinking velocity is not only determined by water salinity, particle density, and size but also by the particles shape leading to considerable differences between measurements and calculated values based on theoretical formulas. Thus, experimental studies are essential to get a basic knowledge about the sinking behavior of microplastics and to provide a representative dataset for model approaches estimating the distribution of microplastics in an aquatic system.

The sinking velocity of microplastics may be altered significantly by weathering and biofouling. Therefore, experimental work will be extended to aged and fouled polymer particles to gain a complete and more realistic documentation of the sinking behaviour.

REFERENCE

Glockzin, M., Pollehne, F., Dellwig, O., 2014. Stationary sinking velocity of authigenic manganese oxides at pelagic redoxclines. Mar. Chem. 160, 67–74.

Pilot Study on Microlitter in the Surface Waters of the Gulf of Finland, Baltic Sea

K. Lind and K. Pullerits
Marine Systems Institute at Tallinn University of Technology, Tallinn, Estonia

For further information on this study, please contact the authors.

Persistent Organic Pollutants Adsorbed on Microplastics From the North Atlantic Gyre

M. Martignac[1], L. Ladirat[1], P. Wong-Wah-Chung[2], P. Doumenq[2], H. Budzin-ski[3], E. Perez[1] and A. ter Halle[1]

[1]Université Paul Sabatier, Toulouse, France [2]Aix-Marseille Université, Aix-en-Provence, France [3]Université de Bordeaux, Bordeaux, France

More than 260 million tons of plastic are used each year. Based on population density and economic status of costal countries the mass of land based plastic waste entering the ocean was recently estimated between 4.8 and 12.7 million metric tons per year (Jambeck et al., 2015). Plastic debris is abundant and widespread in the marine habitat. Marine plastic pollution has been recently recognized as a global environmental threat (Moore, 2008).

Because several plastics are composed of monomers and additives and could adsorb chemical pollutant from aquatic environments, plastic debris found in accumulation zone (gyre) named as microplastics (size <5 mm) also pose a chemical hazard (Teuten et al., 2009). These microplastics are hydrophobic organic material with a lowest density than water, so they are a favorable material for the absorption of persistent organic pollutants (POPs).

Among POPs likely to be adsorbed on these microplastics, polycyclic aromatic hydrocarbons (PAHs) and polychlorinated biphenyls (PCBs) tend to be found rather than dissolved in water because of their hydrophobic nature.

The present study aimed at giving preliminary results about measured concentration of PAHs and PCBs adsorbed on microplastics collected at the surface of the North Atlantic gyre during the sea campaign *"Expédition 7ème Continent"* together with a physicochemical characterization of the microplastics. The results will be also very useful to better understand the biological response to the plastic in terms of transfer of POPs in case of ingestion.

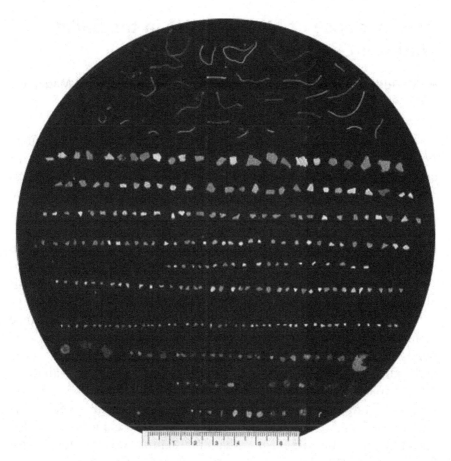

Microplastics collected at the surface of the North Atlantic gyre during the sea campaign "Expédition 7ème Continent."

REFERENCES

Jambeck, et al., 2015. Science 347 (6223), 768–771.

Moore, 2008. Environ. Res. 108 (2), 131–139.

Teuten, et al., 2009. Philos. Trans. R. Soc. B 364, 2027–2045.

First Evidence of Microplastics in the Ballast Water of Commercial Ships

M. Matiddi, A. Tornambè, C. Silvestri, A.M. Cicero and Erika Magaletti

ISPRA, Italian National Institute for Environmental Protection and Research, Roma, Italy

One of the activities of the IPA Adriatic project BALMAS (Ballast Water Management System For Adriatic Sea Protection) is to develop and test ballast water sampling methods and tools for compliance monitoring, in line with the International Convention on the Management of Ships' Ballast Water and Sediments. ISPRA carried out ballast water sampling activity on nine cargo vessels, arrived in the port of Bari (Italy) from July to October 2015. The main scope was to analyze living biological communities (phytoplankton, zooplankton, and indicator microbes) transported in ballast water, but also microplastics were investigated as a side activity, in order to gather some preliminary information on the presence and abundance of microlitter in ballast waters. The ballast water sampling for zooplankton and microplastic analysis was conducted by hand pump and extra strong hose, through tank sounding pipes or manholes. The sampled water was filtered on board using a 50 μm mesh plankton net equipped with removable cod-end. Filtered sample was resuspended in filtered ballast water prior microscopic analysis for detection of viable organisms, than the sample was refiltered with 330 μm mesh and stored in 70% ethanol until microscopic analysis for microlitter (items <5 mm in size) detection. Data showed a huge presence of microplastic particles in the ballast water of each sampled tanks (max: 1410 items m^3, min: 100 items m^3, average 651 ± 160 items m^3), relative to samples collected at sea in other part of the Mediterranean (0.116 items m^2, 0, 15 items m^3). Shape and color composition analysis, showed a higher abundance of synthetic filaments compared to thin plastic layers or fragments (filaments 82%; fragments 7%; layers 11%) while spheres, as plastic virgin pellets, have never been found. Blue was the main color found, followed by black and red items.

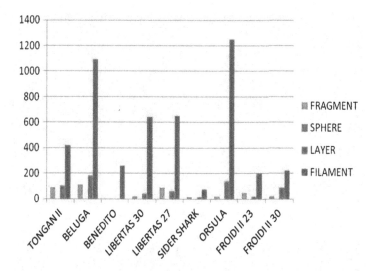

Microlitter Abundance in the Italian Minor Islands, Central Mediterranean Sea

M. Matiddi[1], D. Vani[1], P. Tomassetti[1], A. Camedda[2], G. Zampetti[3],
E. Amato[1], L. Alcaro[1], S. Carpentieri[3], S. Di Vito[3], A. Vianello[4]
and G.A. de Lucia[2]

[1]ISPRA, Italian National Institute for Environmental Protection and Research, Roma, Italy [2]Institute for Coastal Marine Environment and National Research Council (IAMC-CNR), Oristano, Italy [3]LEGAMBIENTE, Roma, Italy [4]Institute for the Dynamics of Environmental Processes and National Research Council (IDPA-CNR), Padova, Italy

Green Schooner is an environmental campaign of analysis and information about Italian sea pollution, promoted and conducted by the Legambiente Onlus every summer since 1986.

In 2015 ISPRA issued an agreement with Legambiente and CNR in order to monitor microlitter in six minor islands, assumed as the clearest Italian water spots, and in two polluted areas. The survey was conducted during the summer edition of the campaign and started with the Po' mouth, the main Italian river, then continued with the archipelago of Tremiti islands, in the southern Adriatic Sea, the Aeolian Islands in the north of Sicily, Ischia island in the Gulf of Naples, Ventotene, one of the Pontine Islands situated in the Central Tyrrhenian Sea and the Tevere river mouth, near Roma. The last sites were the Asinara island, located off the north-western tip of Sardinia and Elba, the largest island of the Tuscan Archipelago.

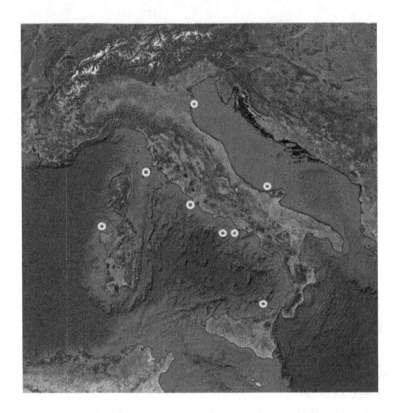

Samples of microplastic and plankton were collected by ISPRA and CNR researchers, using a Manta Trawl and a WP2-FAO, both lined by a 333 μm mesh net. Manta Trawl sampled the top 25 cm of the sea surface while WP2 was forced to work horizontally at a depth around 20 m.

Linear transects were conducted at an average speed of 2 knots for 20 min.

Data showed the presence of microparticles in each sampling area with an average value of 0.3 ± 0.04 items/m^3, Ischia (0.49 items/m^3) resulted the most polluted island. Shape composition analysis, showed a higher abundance of synthetic filaments (50%) followed by fragments (30%), thin plastic layers (17%), and spheres (2%). Compared to the total amount of plankton caught in the samples, abundance of microplastics were low, but their value were higher than many taxa.

A part of the collected microparticles, was analyzed using ATR-FT-IR microspectroscopy in order to identify their chemical composition.

In Search of the Plastic Accumulation Regions: Finetuning Ocean Surface Transport Models

R. Mcadam and E. Van Sebille
Imperial College London, London, United Kingdom

For further information on this study, please contact the authors.

Preliminary Assessment of the Microplastic Presence in the Gulf of Genoa (Italy, Ligurian Sea, Northwestern Mediterranean Sea)

S. Morgana, N. Estévez-Calvar, A. Roveta, R. Stifanese, M. Faimali and F. Garaventa
Istituto di Scienze Marine, Consiglio Nazionale delle Ricerche (ISMAR-CNR), Genova, Italy

Marine litter is mainly composed of plastics that may be physically degraded into smaller debris namely microplastics. Marine microplastics are widely recognized as an emerging concern for marine organisms. Therefore, data about their abundance in the environment are required together with a better understanding of their interaction with marine organisms.

The Mediterranean Sea is known to be a hot spot for floating debris accumulation and, in particular, previous studies highlighted a high density of microplastics in the Ligurian Sea (NW Mediterranean), an area characterized by high anthropogenic pressure.

In this work, surface seawater samples were collected in October 2015 along the Gulf coast of Genoa using a plankton net (80 μm size, 40 cm mouth). Microplastics were sorted under a stereomicroscope, measured, and photographed. Nondestructive Fourier Transform Infrared Spectroscopy (FTIR) was used in order to classify polymers into different families such as polyethylene, polypropylene, and polystyrene. In addition, plankton dry weight was estimated.

The mean abundance of microplastics was $0.13\ \text{item/km}^3$. Polyethylene resulted to be the most represented polymer type. The plankton mass densities resulted to be 10 times higher than plastic densities (plastic:plankton ratio = 0.1).

This work represents a preliminary assessment of microplastics presence in the Gulf of Genoa and confirms the ubiquitous nature of plastic pollution in the Mediterranean Sea.

Further researches are needed in order to clarify the availability for marine organisms and consequently to understand the possible implications along the food web.

Toxicity Assessment of Pollutants Sorbed on Microplastics Using Various Bioassays on Two Fish Cell Lines

P. Pannetier[1], J. Cachot[1], C. Clérandeau[1], K. Van Arkel[2], F. Faure[3], F. de Alencastro[3], F. Sciacca[2] and B. Morin[1]

[1]University of Bordeaux, Talence, France [2]Race For Water Foundation, Lausanne, Switzerland [3]Swiss University, Lausanne, Switzerland

Aquatic ecosystems are subjected to multiple threats including plastic debris. Microplastics, tiny plastic fragments with diameters <5 mm, resulted from runoff and weathering breakdown of larger plastic debris, represent an emerging concern for marine ecosystems. Cosmetic, chemical industry and domestic use including the wastewater of washing machines are additional sources of pollution and accumulation of microplastics particles in aquatic environment. Microplastics impacts on aquatic life are little studied. These small particles could be ingested directly by organisms and cause chronic, physical, and toxicological effects. Moreover, microplastics are the support for a lot of chemicals present in aquatic environment, especially hydrophobic substances. The REACH Regulation and the European directive on the protection of animals used for scientific purposes wish the establishment of alternative to animal experiments. In this concern, toxicological assays on fish cell lines are being developed as alternative methods to provide fast and reliable results on the toxic and ecotoxic properties of chemicals or mixtures. In this aims, rainbow trout liver cell line RTLW-1 and Japanese medaka embryos cell line OLCAB-e3 were used to evaluate toxic effect of water and organic extracts of microplastics artificially coated with B[a]P and PCB 126 to validate different assays e.g., MTT, EROD, and comet assay. In addition, these bioassays were used on both cell lines to analyze the effects of different microplastics samples from Bermuda's and Hawaii's beaches. No toxicity was observed for virgin microplastics whatever the cell line and the bioassay used. Cell lines exposed to microplastics organic extracts from artificially coated particles showed EROD and DNA damage induction in a similar manner as cell lines exposed to the chemical alone. For environmental microplastics, low or no cytotoxicity was observed on both cell lines. However, EROD activity was induced and genotoxicity was observed for certain organic extracts. Difference of sensitivity was observed between both cell lines. Preliminary results obtained here support that cell lines could be an interesting tool to evaluate the potential chemical toxicity of microplastics if chemicals are bioavailable and released in organism.

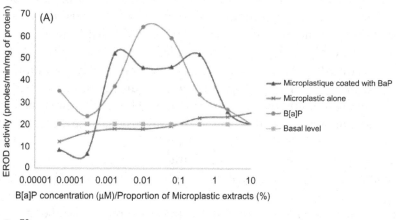

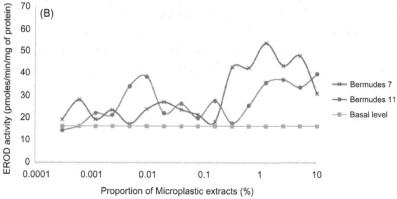

Evolution of EROD activity with proportion of microplastic extracts for artificially coated microplastics and pure chemical (A) and for environmental microplastic samples from Bermuda (B).

Fate of Microplastics in Soft Marine Sediments

P. Näkki, O. Setälä and M. Lehtiniemi

Finnish Environment Institute, Helsinki, Finland

Microplastics are small plastic fragments (<5 mm) present in oceans worldwide. One hypothesis is that most of the plastics will eventually sink to the seafloor, where their interactions with marine organisms and effects in benthic ecosystems are not yet understood. A laboratory experiment was conducted to investigate the role of bioturbation in the localization of the sedimenting microplastics in soft sediments. Pieces of fishing line (three different sizes, <300 μm) were added to 30 sediment containers. Half of these experimental units contained common benthic fauna of the northern Baltic Sea (*Macoma balthica*, *Marenzelleria* spp., *Monoporeia affinis*) in natural densities to model the function of the community, while the other half served as control units without animals. After the experiment the sediment cores were sliced according to depth and the added microplastics in each layer were separated from the sediment using saturated NaCl solution, and counted. The ingested microplastics were counted from dissected animal tissue with epifluorescence microscopy. The results show the ingestion of fishing line by the benthic organisms and the ability of the community to transport microplastics deeper in the sediment. According to the results also the characteristics of microplastics, such as shape and size, may contribute to their localization in the sediment. This study is unique in using microplastics that actually exists in the marine environment and possibly the first to demonstrate the transportation of such microplastics in the sediment by benthic invertebrates. The information on interactions between microplastics and benthic fauna is crucial when assessing the fate and effects of microplastics; distribution and availability to the animals on the seafloor.

Microplastic as a Vector of Chemicals to Fin Whale and Basking Shark in the Mediterranean Sea: A Model-Supported Analysis of Available Data

C. Panti[1], M.C. Fossi[1], M. Baini[1] and A.A. Koelmans[2]
[1]University of Siena, Siena, Italy [2]Wageningen University, Wageningen, The Netherlands

Baleen whales and basking sharks are endangered species that are exposed to microplastic ingestion as a result of their filter-feeding activity. However, the impacts of microplastics on these large marine vertebrates are unknown. Previous studies have provided links between presence of microplastic particles in the Mediterranean, chemical concentrations in neustonic/planktonic and microplastic samples and presence of same chemicals in blubber and tissues of the fin whale (*Balaenoptera physalus*) and the basking shark (*Cetorhinus maximus*). Among these chemicals were plastic-associated chemicals like di-(2-ethylhexyl) phthalate (DEHP) and its primary metabolite, mono-(2-ethyl-hexyl) phthalate (MEHP). Microplastic has been previously detected in the intestines of baleen whales.

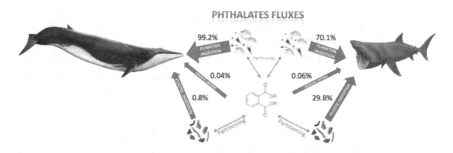

Here we synthesize the available data by providing a model analysis that quantifies the chemical fluxes of plastic-associated chemicals towards large filter feeder marine organisms. A previously published plastic-inclusive bioaccumulation model framework was parameterized to mimic the main biological feeding traits of *B. physalus* and *C. maximus*, i.e., engulfment-filtering and continuous filtering, respectively. We calculated ingestion rates for regular food (krill and plankton), for microplastic, and for the chemicals accumulating via these two pathways, for both species. Aqueous phase chemical concentrations were estimated from chemical concentrations in plankton and by assuming lipid based equilibrium partitioning, after which bioaccumulation via dermal exposure was quantified. The relative shares of chemical uptake via microplastic ingestion, dermal exposure and intake of regular food were assessed and evaluated.

Model analysis revealed that for the fin whale, dermal uptake, plankton ingestion, and plastic ingestion contributed for 0.04, 99.2, and 0.8% to bioaccumulation of MEHP. For the basking shark, these percentages were 0.06, 70.1, and 29.8. This illustrates that the contribution of microplastic ingestion to bioaccumulation is highly dependent on the feeding traits of the species, and that model analysis is a useful tool to understand the underlying mechanisms.

Methodological Prerequisites For Toxicity Testing of Microplastics Using Marine Organisms

J.-W. Park[1,2], N.-H. Hong[1] and Y.-J. Jung[1]
[1]Korea Institute of Toxicology, Daejeon, Republic of Korea [2]Korea University of Science and Technology (UST), Daejoen, Republic of Korea

As increasing worldwide production and usage of plastics, it has been reported that they are ending up and accumulating in ocean, followed by causing harmful effects on marine ecosystems. The plastics can break down into tiny plastics sized from nanometers to micrometers via decomposition processes such as weathering, photo- and biodegradation, which is called secondary microplastics. In recent years, concerns about microplastic pollution have grown among marine toxicologists due to potential toxicity driven from their physical properties such as size (nano to micro) and shape (fibers, irregular), and/or from chemical properties such as polymer compositions (polyethylene, polypropylene, etc.), residual chemicals from manufacture (bisphenol-A, phthalate, etc.) and adsorbed co-contaminants on microplastics. Microplastics could be more toxicologically relevant to organisms than large sized plastics as they can be highly bioavailable (confusing as preys and easily swallowed). Therefore, it is necessary to investigate how microplastics can influence on organisms in marine ecosystem, which requires reliable toxicity testing methods. Generally, maintaining stability of test substances is prerequisite for reproducible and reliable toxicity testing. To precisely evaluate microplastic toxicity, it is critical to rigorously characterize their fate and behavior in testing system, especially for stability during exposing period because microplastics are not soluble and stable in aqueous phase (settling and/or floating), which will influence bioavailability to test organisms and in turn, fail to produce reproducible and reliable toxicity information. Here, physical such as stirring, sonication, and chemical approaches such as artificial and natural surfactants were applied and compared their capabilities to improve aqueous dispersion and maintain stability in seawater. Acute and chronic marine toxicity results of microplastic (polyethylene, 53–63 µm) applied the approach is presented in another poster. This study will help to understand potential risks of microplastics to human and marine environmental health by providing reliable toxicity information in the regulatory point of view.

POPs Adsorbed on Plastic Pellets Collected in the Adriatic Region

P. Makorič and M. Pflieger
University of Nova Gorica, Nova Gorica, Slovenia

In the past years, several studies have revealed the presence of organic contaminants at concentrations from sub ng/g to hundreds g/g on plastic pellets found in coastal environment worldwide. Plastic pellets, categorized as microplastics (<5 mm), are industrial raw material in the shape of small granules (Ogata et al., 2009). They can be unintentionally lost in the environment (e.g., in marine and coastal compartments) during transport and manufacturing. Organic pollutants associated to pellets are either additives (e.g., PBDEs) that are incorporated to plastics during production processes or hydrophobic chemicals which adsorb from the surrounding environment (e.g., seawater). Among these chemicals, some are recognized as persistent organic pollutants. Thus, in order to better assess the impact of plastic pellets in coastal environment, it is necessary to determine the level of adsorbed organic pollutants. The present study was carried out in the frame of DeFishGear project, which focuses on marine litter and microplastics issues in the Adriatic region. This investigation aimed at developing an experimental protocol allowing the quantification of 11 organochlorine pesticides and related degradation products (OCPs), as well as 25 poly-chlorinated biphenyl congeners (PCBs). The plastic pellets, sampled on beaches located in Croatia, Italy, Slovenia, and Greece were first sorted by color. The chemicals were extracted from the plastic matrix in a pressurized fluid extractor (60°C, 100 bar). The extracts were evaporated to 1 mL and then cleaned on Florisil sorbent through solid-phase extraction (SPE). The concentrated and cleaned extracts were quantified on gas chromatography equipped with a microelectron capture detector (GC-ECD). The preliminary investigations revealed the presence of both OCPs and PCBs on the pellets from all sampling sites, with higher concentrations found in yellowish pieces.

ACKNOWLEDGMENTS

The DeFishGear project (http://www.defishgear.net/) is cofunded by the European Union, Instrument for Pre-Accession Assistance (IPA).

REFERENCE

Ogata, Y., Takada, H., Mizukawa, K., et al., 2009. International Pellet Watch: global monitoring of persistent organic pollutants (POPs) in coastal waters. 1. Initial phase data on PCBs, DDTs, and HCHs. Mar. Pollut. Bull. 58, 1437–1446.

The EPHEMARE Project: Ecotoxicological Effects of Microplastics in Marine Ecosystems

F. Regoli, S. Keiter, R. Blust, M. Albentosa, X. Cousin, A. Batel, K. Kopke and R. Beiras
Polytechnic University of Marche (DiSVA-UNIVPM) Via Brecce Bianche, Ancona, Italy

Microplastics (MPs) show the potential to play a remarkable role in the incorporation and trophic transfer of pollutants into the marine food webs but their toxic effect on the marine organisms are still unclear and needs further investigation.

The European Project EPHEMARE (Ecotoxicological Effects of Microplastics in Marine Ecosystems) funded by the JPI Oceans, targets the uptake, tissue distribution, final fate and effects of MPs in organisms representative of pelagic and benthic ecosystems, and the potential role of MPs as vectors of Persistent Pollutants that readily adsorb to their surfaces.

The multidisciplinary consortium constituted by 14 partners from 10 European countries is working to investigate the adsorption of chemicals on MPs, their ingestion, trophic transfer and chemical release, and a wide array of ecotoxicological effects (from transcriptomic to cell damage and organism responses), utilizing standard biological models from bacteria to fish which the Consortium has classified in four different groups: no feeders (class 1), small filter feeders (class 2), filter feeders (class 3), and predators (class 4).

Different typologies of polymers both virgin and previously contaminated and exposure conditions will be tested in order to highlight and validate mechanistic relationships and mode of action of MPs and associated chemical compounds. The main standard materials used will be polyethylene and polyvinyl chloride microparticles but polyethylene terephthalate, polypropylene, and polystyrene microparticles will also be considered for specific experimental approaches.

MPs sizes ranges from 1 to 5000 μm depending what is taken up by each biological model.

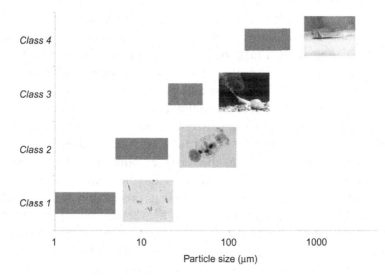

The project research will allow to identify and implement an extensive number of techniques that will assess the damage of MPs in biological processes in the marine environment, from the cellular level to the organism.

Suspended Microsized PVC Particles Impair the Performance and Decrease Survival in the Asian Green Mussel *Perna viridis*

S.E. Rist[1], K. Assidqi[2], N.P. Zamani[2], M. Huhn[2,3] and M. Lenz[3]

[1]Technical University of Denmark, Kongens Lyngby, Denmark [2]Bogor Agricultural University, Bogor, Indonesia [3]GEOMAR Helmholtz Centre for Ocean Research Kiel, Kiel, Germany

The majority of marine microplastics will finally sink to the seafloor where the particles become available for a large variety of benthic species. Due to their feeding modes, benthic invertebrates are presumably particularly affected by microsized particles. However, the number of studies that investigated the effects of plastic pollution on seafloor organisms so far is very limited. Furthermore, the experiments done mainly applied short exposure periods (several hours to days), what limits their ecological interpretability. We conducted a laboratory exposure experiment with the Asian green mussel *Perna viridis* for 91 days. To investigate the relevance of a so far neglected exposure pathway, we mimicked the resuspension of plastic particles from the sediment by tidal currents. Furthermore, a wide range of polyvinylchloride particle concentrations (0 mg/L, 21.6 mg/L, 216 mg/L, and 2160 mg/L) was chosen to identify the threshold beyond which an impact on the physiological performance of the animals becomes detectable. Mussels were exposed to these particle concentrations during two resuspension events per day of which each lasted for 2 h. After resuspension, which was induced by an air stream, the particles sank down to the bottom of the experimental containers. After 44 days of exposure, mussel filtration rates, respiration rates, and byssus production were found to decrease with increasing particle loads. Moreover, during the 91 days of the experiment mussel survival declined with increasing particle concentration. These negative effects presumably go back to prolonged periods of valve closure that limited food uptake and reduced the energy budget that was available for maintaining the mussels' metabolism.

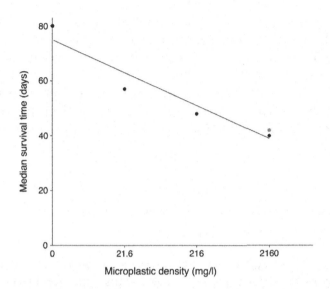

Priority Pollutants in Microplastics From Beaches in Gran Canaria Island

M.R. Sanz, V.M. Gabet and J.R.B. Rodríguez
Instituto Tecnológico de Canarias, Las Palmas, Spain

For further information on this study, please contact the authors.

Plastic Debris in Mediterranean Seabirds

**A. Rodríguez, T. Militão, J. Martínez, M. Codina-García
and J. González-Solís**
Universitat de Barcelona, Barcelona, Spain

Plastic debris is usually ingested by seabirds by mistake or because plastic fragments resemble their natural food items. After ingestion, plastics can produce intoxication, internal wounds, digestive tract blockage, and ulcers among other affections.

Therefore, there is an urgent need to evaluate the extent of plastic ingestion in seabirds. In the present study we attempted to understand its importance in Mediterranean seabirds and to evaluate the differences among species. We quantified, classified, measured, and weighted the plastics accumulated in 316 stomachs from 3 shearwater and 2 gull species accidentally captured by longliners in the western Mediterranean from 2003 to 2015. Scopoli's shearwaters (*Calonectris diomedea*) showed the highest plastic frequency (71.58%), followed by Balearic shearwaters (*Puffinus mauretanicus*, 46.91%) and Yelkouan shearwaters (*Puffinus yelkouan*, 56.86%). In contrast, the gulls presented the lowest.

Monitoring Plastic Ingestion in Selected Azorean Marine Organisms

Y. Rodríguez[1], J.P.G.L. Frias[2,3], R. Carriço[2,3], V. Neves[2,3,4], J. Bried[2,3], H.R. Martins[2,3], F. Vandeperre[2,3], M.R. Santos[5], J.A. van Franeker[6], A.B. Bolten[7], K.A. Bjorndal[7] and C.K. Pham[2,3]

[1]OMA – Observatório do Mar dos Açores, Horta, Portugal [2]MARE – Marine and Environmental Sciences Centre, Horta, Portugal [3]IMAR – Instituto do Mar, Horta, Portugal [4]University of the Azores, Horta, Portugal [5]DRAM – Direção Regional dos Assuntos do Mar, Horta, Portugal [6]IMARES – Wageningen-UR, Den Helder, Netherlands [7]University of Florida, Gainesville, FL, United States

Despite its geographic isolation from large population centers, the Azores archipelago (north-eastern Atlantic) is not immune to the growing environmental threat of marine litter. Recent research developments suggest that many organisms are directly affected by this issue that should be addressed with consistent monitoring efforts.

The goal of this study is to present the results of recent and past monitoring efforts on plastic ingestion in different food-web components (seabirds, fish, and sea turtles) in the Azores.

A total of 421 dead Cory's shearwaters (*Calonectris borealis*) were collected during different fledgling periods between 2000–12 and 2015 from different islands. These birds were mainly road kills but also included individuals, which collided with buildings and other structures, or birds, which were dehydrated. Plastic fragments were found in the stomachs (proventriculus and gizzard) of a large majority of the fledglings (93%), although quantities were well below levels observed in fulmars in the North Sea.

Plastic ingestion was also monitored in 13 fish species ($n = 209$) of contrasting ecology, ranging from deep benthic to pelagic species. So far, no plastic particles or other anthropogenic debris were found from the sampled digestive tracts.

Preliminary results concerning plastic ingestion by other marine organisms will be provided, as this is an ongoing task of the project. Our results provide important data for establishing appropriate indicator species to monitor ingestion and abundance of plastic in the Azorean waters. The establishment of a regular program to monitor the ingestion of plastic in marine organisms is vital to understand abundance trends and effectively assess future management actions.

Catching a Glimpse of the Lack of Harmonization Regarding Techniques of Extraction of Microplastics in Marine Sediments

E. Rojo-Nieto, J. Sánchez-Nieva and J.A. Perales
University of Cadiz, Cadiz, Spain

Plastic is the most abundant marine litter category, and due to its characteristics can remain in the environment for long periods of time, eventually causing impacts on wildlife, tourism, fishing, and shipping activities (Arthur et al., 2009; Andrady, 2011; Frias et al., 2016; Ivar do Sul and Costa, 2014; Jang et al., 2014; Thompson et al., 2004). Microplastics are well-documented pollutants in the marine environment that result from fragmentation of larger plastic items (Frias et al., 2016). The ubiquitous presence and persistency of microplastics in aquatic environments are of particular concern since they represent an increasing threat to marine organisms and ecosystems (Phuong et al., 2016).

At present, there is no universally accepted definition regarding the size of microplastics (Van Cauwenberghe et al., 2015). Nowadays, most researchers agree with the definition of MPs proposed by Arthur et al. (2009) of microplastics as particles in a size range of less than 5 mm. Marine litter is addressed for the first time in Marine Strategy Framework Directive (MSFD) in an integrated way towards the protection of the marine environment (Frias et al., 2016; Galgani et al., 2013).

Recently stated that microplastics such as pellets have been reported for many years on sandy beaches around the globe. Nevertheless, high variability is observed in their estimates and distribution patterns across the beach environment are still to be unraveled. In this work, a review regarding microplastic studies has been conducted. The main objective of this review is to assess the state of the science in the determination of microplastics in the marine environment, more specifically in marine coasts and beaches. This review has shown that there is a clear need for standardized techniques and unified results reporting. For this reason, in this work we have mainly focussed on extracting techniques, type of plastics studied and recoveries obtained in these studies. In addition, proposals for further studies to promote an extraction procedure harmonization have been added.

REFERENCES

Andrady, A.L., 2011. Microplastics in the marine environment. Mar. Pollut. Bull. 62 (8), 1596–1605.

Arthur, C., Baker, J., Bamford, H. (Eds.), 2009. Proceedings of the International Research Workshop on the Occurrence, Effects and Fate of Microplastic Marine Debris, Sept 9–11, 2008. NOAA Technical Memorandum NOS-OR&R-30.

Frias, J.P.G.L., Gago, J., Otero, V., Sobral, P., 2016. Microplastics in coastal sediments from Southern Portuguese shelf waters. Mar. Environ. Res. 114, 24–30.

Galgani, F., Hanke, G., Werner, S., De Vrees, L., 2013. Marine litter within the European Marine strategy framework directive. ICES J. Mar. Sci. 70, 1055–1064.

Ivar do Sul, J.A., Costa, M.F., 2014. The present and future of microplastic pollution in the marine environment. Environ. Pollut. 185, 352–364.

Jang, Y.C., Hong, S., Lee, J., Lee, M.J., Shim, W.J., 2014. Estimation of lost tourism revenue in Geoje Island from the 2011 marine debris pollution event in South Korea. Mar. Pollut. Bull. 81 (1), 49–54.

Phuong, N.N., Zalouk-Vergnoux, A., Poirier, L., Kamari, A., Châtel, A., Mouneyrac, C., et al., 2016. Is there any con-sistency between the microplastics found in the field and those used in laboratory experiments?. Environ. Pollut. 30 (211), 111–123.

Thompson, R.C., Olsen, Y., Mitchell, R.P., Davis, A., Rowland, S.J., John, A.W.G., et al., 2004. Lost at sea: where is all the plastic? Science 304 (5672), 838.

Van Cauwenberghe, L., Claessens, M., Vandegehuchte, M.B., Janssen, C.R., 2015. Microplastics are taken up by mussels (Mytilus edulis) and lugworms (Arenicola marina) living in natural habitats. Environ. Pollut. 199, 10–17.

Floating Plastics in the Sea: People's Perception in the Majorca Island (Spain)

L.F. Ruiz-Orejón[1], J. Ramis-Pujol[2] and R. Sardá[1]
[1]Centre d'Estudis Avançats de Blanes (ceab), Girona, Spain [2]ESADE Barcelona, Barcelona, Spain

Plastic pollution problems have been neglected in marine research until recently. Today, major awareness of the problem by society and new knowledge in the field make relevant to investigate about perceptions, attitudes, and behavior of people. This study is aimed to contribute to the approach of the social dimension of problem on two (qualitative and quantitative) studies carried out in the Island of Majorca during 2015. For this purpose, they were convened three representative focus groups of Majorca's stakeholders (Science-NGO, Public Administration, and Companies) in conjunction with a beach survey to collect citizen' insight ($N = 635$). In terms of responses, study participants prioritized excess plastic and the lack of awareness as the main drivers of the problem, suggesting a worst situation in the future. These insights may be considered for research and practices of stakeholders' engagement on plastic pollution derived consequences.

Plastics in the Mediterranean Sea Surface: From Regional to Local Scale

L.F. Ruiz-Orejón[1], R. Sardá[1] and J. Ramis-Pujol[2]
[1]Centre d'Estudis Avançats de Blanes (ceab), Girona, Spain [2]ESADE Barcelona – Sant Cugat, Barcelona, Spain

Using two sea voyages throughout the Mediterranean (2011 and 2013) and four seasonal campaigns carried out in the Channel between Majorca and Menorca in 2014 and 2015, a total of 139 samples for floating plastic debris were obtained with a Manta trawl (rectangular opening of 1.0×0.25 m and $333\,\mu$m mesh). In the Mediterranean Sea floating plastic weight concentration in samples averaged 579.9 g dw/km^2 with a maximum value of 9298.2 g dw/km^2. Particle density concentration for the sizes collected yielded an average of 147,500 pieces/km^2 with a maximum value of 1,164,403 pieces/km^2. A general estimate totaled a value of 1455 tons dw of floating plastic in the region. Obtained plastic size distribution for sampled sizes showed a kind of surprising cohort-type distribution with several sizes more frequent than others with a repeated pattern. The characteristics of the Mediterranean Sea make it a high risk area to accumulate plastics. In the Majorca–Menorca Channel some seasonality in the accumulation of plastics was observed, during spring and summer plastics were accumulated in the southwestern area while during autumn and winter distribution was more homogeneous through the investigated area.

Analysis of Organic Pollutants in Microplastics

S. Santana-Viera, R. Guedes-Alonso, C. Afonso-Olivares, S. Montesdeoca-Esponda, M.E. Torres-Padrón, Z. Sosa-Ferrera and J.J. Santana-Rodríguez
Universidad de Las Palmas de Gran Canaria, Las Palmas de Gran Canaria, Spain

For further information on this study, please contact the authors.

From the Sea to the Dining Table and Back to the Environment: Microlitter Load of Common Salts

O. Setälä[1], A. Koistinen[2], S. Budimir[1], S. Hartikainen[2], M. Lehtiniemi[1], P. Näkki[1], M. Selenius[2] and J. Talvitie[3]
[1]Finnish Environment Institute, Helsinki, Finland [2]University of Eastern Finland (UEF), Kuopio, Finland [3]Aalto University, Aalto, Finland

Microplastics are widespread in marine environment from the coastlines to the open ocean. Salts originated from the sea may concentrate microplastics from large volumes of seawater and carry them back to land. In addition to the direct human consumption of table salts, various types of salts are also used for example in agriculture as preservatives and to complement animal nutrition. Furthermore, salts are widely spread on roads during winter to defrost snow and ice in the northern Europe. We analyzed and compared common sea salts, mineral salts and mountain salts, which are available in Finnish supermarkets and manufactured around the world. Salts were dissolved in water and filtered through plankton nets (mesh size 300, 100, 50, and 20 μm). Particles caught on filters were examined using FTIR microscopy to characterize the material of the particles. Preliminary results show that all analyzed salt products contained microlitter, and most of these particles were fibers, including both organic and polymer fibers. This study reveals a potential pathway of microplastics relocalization from one area to another and between marine and terrestrial environments.

Sandy Beaches Microplastics of the Crimea Black Sea Coast

S. Elena and C. Igor

The A. O. Kovalevsky Institute of Marine Biological Research of RAS, Sevastopol, Russian Federation

Microplastics abundance was estimated from surveys on two of the most popular Sevastopol sandy beaches of the Crimea Black Sea Coast (Omega beach and Uchkuevka beach). The samples were collected during February–March 2016 from the top 5 cm of the numerous square areas (1 × 1 m) placed on 20 m long transects along the shore line. Three type of stainless steel sieves were used: mesh sizes 5, 1, and 0.3 mm. In the laboratory, the collected sediments were introduced into a glass tank with a high concentration solution of sodium chloride (NaCl) 140 g/L, the floating plastic particles recovered, sorted, and categorized by type, usage, and origin.

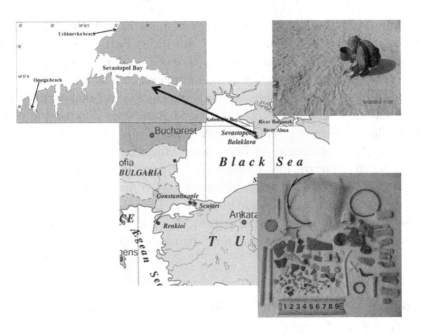

Study area with sampling sites.

The mean microplastics densities on Omega and Uchkuevka beach were 4.2 and 2.6 items/m^2, accordingly. Also we estimated macroplastic particles abundance from the same investigated square areas. The monitoring will be continued on other Crimea Black Sea Coast beaches in different seasons to quantify and qualify coastal litter pollution and develop effective mitigation measures.

Plastic Prey; Are Fish Postlarval Stages Ingesting Plastic in Their Natural Environment?

M.K. Steer[1,2], P.K. Lindeque[1] and R.C. Thompson[2]
[1]Plymouth Marine Laboratory, Plymouth, United Kingdom [2]University of Plymouth, Plymouth, United Kingdom

For further information on this study, please contact the authors.

Microplastic Ingestion by Planktivorous Fishes in the Canary Current

A. Štindlová, P. Garrido, A. Herrera and M. Gómez
Universidad de Las Palmas de Gran Canaria, Las Palmas de Gran Canaria, Spain

Marine plastic debris is present worldwide across the ocean water column and can affect marine habitat and the organisms living in it. In the last decade, microplastics (MPs) have become a subject of intense investigations due to the increasing concerns about their negative impact on wildlife and possible toxicity on living organisms (including humans). In the ocean MPs can be easily ingested by numerous marine organisms because of their small size (<5 mm). The Northwest African upwelling system is an important fishery region. This study is the first one to reveal the presence of MP particles in the stomach contents of two zooplanktivorous fish species in the Canary Current region: bogue (*Boops boops*) and Atlantic chub mackerel (*Scomber colias*). From 64 fish samples examined, 21 fish stomachs (33%) contained microplastic fragments.

First Quantification of Microplastic in Norwegian Fjords Through Nondisruptive Ad-hoc Sampling

M.S. Nerheim[1] and A. Lusher[2,3]
[1]University of Bergen (UIB), Bergen, Norway [2]Galway-Mayo Institute of Technology (GMIT), Galway, Ireland [3]National University of Ireland-Galway (NUIG), Galway, Ireland

The Norwegian fjord systems are important grounds for recreational and industrial use, as well ecosystem services. The biodiversity of Norwegian fjords have been and are being extensively studied, often due to their importance for fishing and aquaculture industries. Despite the worldwide studies on the threats from microplastic to a large range of species and ecosystems, no studies or monitoring efforts have looked at the quantity of microplastic in Norwegian fjord systems.

This study presents the first quantification of microplastic in Norwegian fjords, including a preliminary analysis of distribution effects, and clearly shows the presence of microplastic in Norwegian fjords. Additionally we show how microplastic sampling can easily be included as a nondisruptive ad-hoc monitoring strategy on board a large variety of vessels, including research vessels, commercial freight and transport, and recreational vessels.

Further development and implementation of this method in terms of sampling and monitoring will allow for easy quantification and it can be easily adapted for worldwide implementation.

Abundance of Microplastics and Adhered Contaminants in the North Atlantic Ocean

K. Syberg[1], C.M.H. Knudsen[1], Z. Tairova[2], F. Khan[1], Y. Shashoua[3], T. Geertz[4], H.B. Pedersen[4], C. Sick[4], T.B. Knudsen[1], J.H. Jørgensen[5] and A. Palmqvist[1]

[1]Roskilde University, Roskilde, Denmark [2]Aarhus Universitet, Aarhus C, Denmark [3]The National Museum of Denmark, København K, Denmark [4]Plastic Change, Hellerup, Denmark [5]NIRAS A/S, Allerød, Denmark

For further information on this study, please contact the authors.

Microbes and Bioplastic

T. Valentina and T. Tinkara

National Institute of Biology, Piran, Slovenia

For further information on this study, please contact the authors.

Marine Litter Monitoring for Coastal Management Indicator System Development: Citizen Science and Collaboration Communication Approach

U. Janis, E. Raimonds, G. Sintija, K. Janis and B. Janis

University of Latvia, Riga, Latvia

Marine litter in-depth monitoring programs now being only at the early stages of development do place high importance on the evaluation of beach litter in Latvia. Beach litter monitoring has been elaborated on the data collected and provided by the campaign "My Sea" since 2012. One of the broadest public monitoring programs and real citizen science activity is to be more integrated into both science programs and coastal planning also at the local municipal level particularly, what is also the aim of EU BONUS research program funded project BaltCoast "A Systems Approach Framework for Coastal Research and Management in the Baltic" (cofunding this report).

Marine litter surveys at Salacgriva municipality coastline (BaltCoast case study site) have been done in four monitoring fields set upon according to criteria stated out in UNEP and OSPAR methodologies and representing also the diversity of Salacgriva coastline. The average amount of beach litter in Salacgriva is 205 items/100 m, according to UNEP methodology, what is above the average amount—173 items/100 m on the Latvian coast. Also, taking into account that provisional target of good environmental status is 135 items/100 m, Salacgriva county is given status descriptor—noncompliant marine/beach litter status with negative trend. Beach litter data also exposes sources and the fractions of litter, thus showing that more than 52% of litter are artificial polymers and 25% of all litter are from sea-based sources. The land based litter problem in Salacgriva is related first of all to the tourist used beaches, while not so extensively used beaches are well on track in reaching good environmental status, averaging almost three times less litter items. Local population perceptions and understanding sea-based problems are very diverse and fragmented, what also additionally stresses the need for improvements in coastal planning.

Coastal sustainable governance at local municipalities require application of corresponding indicator systems (IS), even traditional existing IS are nonrelevant for municipality level because of their spatial scale, orientation on action results, and very limited attribution to the coast and even less to the marine environment. IS shall be based on nature and social sciences studies, and especially on their interactions as for the coastal socioecological systems. IS are to be developed also for both top-down and bottom-up complementary approaches, including particularly citizen science and communication developments. Relevant IS models are to be tested correspondingly applying collaboration communication model by involving all stakeholders and both still not complementary sea- and land based monitoring, improving sea-coast governance decision-making processes.

Types and Concentration of Microplastics Found on Remote Island Beaches During the Race for Water Odyssey.

C. Levasseur[1], F. Faure[1], K. van Arkel[2], F. Sciacca[2] and L.F. de Alencastro[1]

[1]Swiss Federal Institute of Technology, Lausanne, Switzerland [2]Race For Water Foundation (R4W) Lausanne, Switzerland

In 2015, The Race for Water Foundation, a NGO dedicated to water preservation, conducted the Race For Water Odyssey.

This Odyssey was a unique environmental expedition, which sailed across the Atlantic, Pacific, and Indian Oceans to compile a human and environmental assessment of the current state of plastic pollution in the oceans.

Three main goals were followed: study plastic pollution on beaches of remote islands located within the oceanic gyres (which act as a sort of natural barrier against the movement of marine debris), raise awareness on the issue and identify solutions preventing waste from ending in the ocean.

The team realized seven scientific island stopovers (Azores, Bermuda, Easter Island, Hawaii, Guam, Koror, and Rodrigues) and went to seven main coastal cities (Bordeaux, New York, Valparaiso, Tokyo, Shanghai, Cape Town, Rio de Janeiro, and the arrival in Bordeaux). Due to boat capsizing in the Indian Ocean, local scientists helped us finalizing the sampling on Chagos and Tristan da Cunha, using the same protocol.

At each stopover, the National Oceanic and Atmospheric Administration's (NOAA) standardized protocol has been performed to identify and quantify macrodebris, and collect microplastic samples on shorelines. Microplastic samples are currently being analyzed in three laboratories to categorize them and assess their concentration, to analyze the ecotoxicological effect on fish larvae, and the persistent organic pollutants adsorbed on their surface.

Plastics have been found on every beach during the Odyssey at various concentrations. The worst spot was in Hawaii, Kamilo beach, with a maximum density of 3357 macroplastics (i.e., items larger than 2.5 cm) by 100 square meters, and more than 30,000 microplastics (i.e., items smaller than 0.5 cm) on average per square meter.

Here, we will focus more on preliminary results of the Odyssey, specifically on microplastics concentration and typology according to the categories (hard, foam or film fragments, fibers, microbeads, pellets...) and polymer types (PET, HDPE, LDPE, PVC...) collected during this expedition.

A Throwaway Society: Is Science Stuck With Single Use Plastic? What Can We Do About It?

A. Watts[1], E. Reardon[1] and M. Urbina[1,2]

[1]University of Exeter, Exeter, United Kingdom [2]Universidad de Concepción, Concepción, Chile

Society is starting to change in terms of single use plastics: plastic bag charges, banning microbeads in cosmetics to name just two. We are finding political, behavioral, and technological solutions to reduce the use of single use plastics: turning our backs on the throwaway society. We are however left with an issue. If you are a scientist based in a laboratory, aquarium or field based how much plastic do you use to deliver science? Can we reduce this amount? Is there something that the science community should do as a whole to reduce this? These were just a few starting questions. We have calculated that 280 bench scientists at the University of Exeter produced 267 tons of plastic waste in 2014, when extrapolated to the 20,500 institutions worldwide this amounts to 5.5 million tons of plastic going to waste or 1.8% of total worldwide plastic production. Values are published in Nature 2015, 528(7583).

The aim of this poster is to highlight the issue of plastic waste generated in research activities to the minds of active researchers and to enable an open discussion on the opportunities and barriers to reducing our dependency of laboratory based single use plastic. Please visit the poster and add your suggestions or tweet suggestions to it using the hashtag #plasticfreescience.

Linking Education and Science to Increase Awareness of Marine Plastic Litter— Distribution of Plastic Waste on Beaches of the German Bight

A. Wichels, B. Harth and G. Gerdts
Alfred Wegener Institute Helmholtz Zentrum für Polar und Meeresforschung (AWI), Helgoland, Germany

Since the 1950s more than 6.1 billion tons of plastics have been produced. It has been estimated that about 10% of this amount will be deposited long-term in the oceans. The problem is highlighted by several current studies using different environmental sampling protocols and analytical methods. Effects of this anthropogenic litter on the environment and organisms are heavily on debate, emphasizing the need to transfer this knowledge to young educated people and to fuel educational programs. The school lab OPENSEA at the Alfred-Wegener-Institute on Helgoland started a joint high school project on marine plastic litter in cooperation with the experts of the marine microplastics group at AWI to link science and education more closely. Based on the OSPAR protocol for beach monitoring of marine litter we developed an experimental set up focusing on sampling and identification of plastic litter on beaches, shores and in sediments in the course of the OPENSEA science and education program for grammar and high school scholars.

This monitoring provides environmental data on marine plastic litter and will be integrated in a long term data monitoring program in the course of a citizen science study. In addition we plan to integrate also smaller plastic particles into the project. Fractionated sediment samples will be screened for particles >5 mm, which then will be analyzed by ATR-FT-IR. These educational activities, with a strong link to the latest science and to sophisticated technology will raise the awareness of younger people for the marine litter problematic. We aim at increasing their concerns after taking part in this program. We will present background information, sampling strategies, identification efforts, and results based on this scholar science project.

Are Smaller Microplastics Underestimated? Comparing Anthropogenic Debris Collected With Different Mesh Sizes

A.W. McNeal[1], M. Cole[1,2], T.S. Galloway[2], C. Lewis[2], A. Watts[2], S. Wright[2], R.Z. Miller[3] and P. Lindeque[1]

[1]Plymouth Marine Laboratory, Plymouth, United Kingdom [2]University of Exeter, Exeter, United Kingdom [3]Rozalia Project, Granville, VT, United States

For further information on this study, please contact the authors.

Precipitation/Flotation Effect of Coagulant to Microplastics in Water

M. Yurtsever[1] and İ. Çelik[1]
Sakarya University, Sakarya, Turkey

The environment hazards of plastics used widely and excessively throughout the globe, in the form of pollution made even worse by the fact that it takes virtually centuries for them to break down in the nature are no longer news to anyone. Yet, perhaps the even larger threat plastics would pose for water sources and environment, when broken down into micron-scale particles is only recently getting a place on the agenda. Microplastics are capable of moving around through passive movements in water, as well as adsorbing hazardous chemicals in aquatic environments. They are also mistaken by aquatic creatures, and often swallowed as food. The primary reason deceiving creatures to take plastics as food is the ability of microorganisms to take hold on and grow on plastics, creating in time a natural layer to cover and disguise it. A piece of plastic coated with biofilm can easily look like a piece of "organic matter." According to researchers, the conventional wastewater treatment plants in our cities fail to filter out the microplastics in the sewage, and duly discharge them to the receiving environment.

This study is based on an experimental investigation of how plastics in aquatic environments would react upon the introduction of coagulants used in the treatment of water. The coagulant used in the study was Iron(III) chloride (FeCl3). Experiments using various doses of ferric chloride focused on the behavior of polyvinyl chloride (PVC) microplastics at a scale of approximately 150 μm, with a density of 1.37 g/cm^3, at the water-wastewater treatment plants (WWTP). Two major factors with a defining impact on precipitation experiments were temperature and pH values. The experiments were ran at 20°C and around pH 7.

The experiments were essentially observations of the behavior of a total of 1 g of PVC particles in 1 L of water in a 50 cm high glass column, in response to the addition of coagulants to water. Samples taken from the bottom of the column, as well as heights of 2, 11 cm, just below the surface (20 cm depth), and the surface (22 cm) were then filtered through filter paper with the pore size of 0.45 μm, whereupon PVC microplastic particles in each square section were counted with the use of a microscope. To ensure the accuracy of the count, each sample were counted for a total of 10 times, and the average of all counts was used in the study. The results indicate that the higher the amount of coagulants, the more PVC precipitation in the bottom. Yet, the case with the surface does not follow this rule of thumb. After reaching an optimal level, any further increase in coagulant levels were coupled with increase in PVC volumes observed on the surface. In other words, excessive coagulant doses make PVC particles float on the surface. This analysis represents only the preliminary work in the area. Experiments that are more comprehensive will follow using colloidal plastic (10^{-3} μm < d < 0.1 μm) particles as well.

Personal Care and Cosmetics Products (PC-CPs): Is It Cleaning or Pollution?

M. Yurtsever and U. Yurtsever
Sakarya University, Sakarya, Turkey

This study focuses on microfibers from wet wipes and microbeads from cosmetics—two sources, among others, of microplastics delivered to wastewater treatment plants via sewage—created through personal use at homes. Irresponsible household consumption of personal care products are discussed in terms of their environmental impact, the hazards they pose for aquatic creatures, and the operational problems at environment related facilities, leading to the conclusion that such excessive use would in time bring about a major environmental catastrophe.

It would not be farfetched to call "clean and innocent" wet wipes, which in recent decades found extensive use all around us, "dirty" in terms of their overall impact. The waste produced through their use is by no means "innocent." Given their ease of handling and use, wet wipes came to be a major element of our daily life, and are especially prominent in food service, air, land, and marine transportation, as well as catering and restaurant business. Initially invented in the 1950s, wet wipes entered into wide-spread use from 1990s on. Furthermore, the recent years saw an overwhelming increase in their consumption, given the developments to make wet wipes produced in various forms and quantities affordable. The trend, if unchecked, is set to become a true "calamity" for the environment. "Wet wipe piles" are beginning to appear in treatment plants all around the world, posing substantial problems regarding their operations. Furthermore, they are also ranking high among the waste inventory collected at the beaches, and hence have the potential to find their way to seas and oceans.

However, this study investigates the personal care products containing "microplastics" and being sold in supermarkets in Turkey. The microbeads found in the personal care products thus marketed and used are polyethylene derivatives. The size of microbeads used in personal care and cosmetics products is approximately $1\ \mu m - 1$ mm. The results indicate that men and women who took part in the surveys has yet to develop an awareness about the existence, let alone harms, of microplastics as pollutants. Cosmetics containing microbeads, such as face cleaning gels, peeling gels, shower gels, shampoos, soaps, toothpaste, eyeliner, mascara, lip-gloss, deodorants, and sun lotions are consumed in overwhelming quantities. In addition to an excessive level of use in the cosmetics sector, plastic microbeads are used as corrosive particles in detergents, stainless steel surface cleaners, and cleaning liquids.

Detection of Microplastics with Stimulated Raman Scattering (SRS) Microscopy

L. Zada[1], H.A. Leslie[1], A.D. Vethaak[1,2], J.F. de Boer[1] and F. Ariese[1]

[1]VU Amsterdam, Amsterdam, The Netherlands [2]Deltares, Marine and Coastal Systems, Delft, The Netherlands

For further information on this study, please contact the authors.

PART III

Lanzarote Declaration, June 21 2016

The Lanzarote Declaration summarizes the highlights from the work shared at MICRO 2016, and was produced collaboratively by the members of the MICRO 2016 Organizing Board and Scientific Committee. It represents the first milestone in the process of building the road to MICRO 2018.

Lanzarote Declaration, June 21 2016

From May 25–27, the MICRO 2016 international conference on microplastics was held in the Cabildo de Lanzarote, UNESCO Biosphere reserve of Lanzarote, Canary Islands, Spain. Rooted in the MICRO 2014 international workshop in Plouzané, France and the MICRO 2015 seminar in Piran, Slovenia, MICRO 2016 provided an opportunity to share available knowledge, fill in gaps, identify new questions and engage the scientific community through the work presented and the **Lanzarote Declaration**.

We recognize that the **Lanzarote Declaration** stems from previous regional, national, and international efforts such as: The London Convention (1972); the Barcelona Convention (1976); the MARPOL Convention (1978); the East Asian Seas Action Plan (1981); the Abidjan Convention (1984); the Cartagena Convention (1986); Bâle Convention (1989); the OSPAR Convention (1992/1998/2002/2005/2006/2007); the Northwest Pacific Action Plan (1994); the Nairobi Convention (1996); EU Water Framework Directive (2000); the Teheran Convention (2003); EU Marine Strategy Framework Directive (2008); the Honolulu Commitment (2011); the Manila Declaration (2012); the Mediterranean Regional Plan on Marine Litter (2014); and the G7 Leaders' Declaration (2015).

Nearly all aspects of our daily lives involve plastics. Plastics are versatile, light, durable, inexpensive, and can be shaped to almost any form imaginable. While these are valuable traits, the "disposable" use of plastics in recent decades is now clearly visible in the majority of Earth's ecosystems. Plastics have been found in the atmosphere, soils, fresh water, oceans, seas, and polar regions. They are even recognized as new habitat for organisms, called the Plastisphere. As they become increasingly prevalent in ecosystems, concerns about plastics are mounting due to their unknown effects at the organismal level and potential consequences for ecosystem functioning. Most plastics are considered persistent material and accumulate in the environment since they cannot be mineralized. Over time we find increasing numbers of fragments of decreasing size.

Microplastics are generally defined as any plastic particles <5 mm and they come from two sources: (1) primary microplastics, which

MICRO 2016. Fate and Impact of Microplastics in Marine Ecosystems.

include industrial abrasives, exfoliants, cosmetics, and preproduction plastic pellets; and (2) secondary microplastics, which come from the degradation of larger processed plastic items.

While the presence of microplastics in ecosystems has been reported in the scientific literature since the 1970s, many pressing questions regarding their impacts remain unresolved.

We, the 46 members of the Scientific Committee, sign the **Lanzarote Declaration** on behalf of 632 researchers whose work comprised over 200 presentations at the MICRO 2016 conference. Drawing from the shared scientific and technical material, in this declaration we summarize the highlights from MICRO 2016 and mark the first milestone of the **Road to MICRO 2018** collaborative process.

Highlights from the MICRO 2016 conference:

This declaration covers any type of microplastic.

There is a need to maintain and improve the link between ongoing research and policy efforts at national and international levels such as the EU Marine Strategy Framework Directive, OSPAR, NOWPAP, MEDPOL, etc.

Microplastics are found nearly everywhere that has been investigated in the world's oceans and coastal areas, including the most remote parts of the earth. Though less studied, they are also found in fresh water bodies and terrestrial environments.

The widespread occurrence of microplastics and their impacts have been demonstrated by more than 50 studies worldwide.

As demonstrated by several studies, microplastics may be ingested by many species, and the risk of transfer to humans has been shown for some commercial species such as fish, mollusks, and crustaceans. Seaweed has also been shown as a vector for microplastics.

Studies show the overlap of aquatic biota feeding grounds and waters with high levels of microplastic pollution. This has particularly been demonstrated for: fin whales in the northwestern Mediterranean Sea, marine mammals stranded in Ireland, turtles in Northern Cyprus, turtles in the Canary Current, and seabirds.

Runoff has been robustly demonstrated as a significant microplastic vector, particularly road runoff in populated areas. Several studies confirm sewage sludge as a vector of microplastic pollution, highlighting the need for further studies, and actions focused on sewage treatment plants and the urban water cycle.

From previous studies finding microbial communities on plastic surfaces, new results confirm plastisphere microbial communities and provide site-specific insight into these communities and their successional changes over time.

Modeling is clearly a fundamental and complementary tool for identifying microplastic sources, distribution paths, and potential sinks.

Recent studies have confirmed that large plastic items and microplastics can be further degraded into nanosized plastic particles, which may impact the biosphere. The largely unstudied field of nanoplastic pollution will potentially be of significant importance in years to come.

In order to integrate data across various studies and ongoing projects, we must: (1) standardize the identification and quantification of microplastics, and (2) explicitly describe the techniques and methods currently used in ongoing, nonstandardized studies. There is also a clear need to standardize and harmonize approaches for professional and citizen science efforts, keeping in mind the importance of documenting the benefits of citizen science and the need for standardized databases and interfaces to share the results of citizen science work.

Citizen science contributes to microplastics sampling and monitoring. Outreach and education efforts to raise awareness about microplastics in marine environments and increase ocean and plastic literacy help connect the general public with the issue of microplastics. Perceptions and representations can be changed through science communication.

Working to prevent and mitigate microplastic pollution provides co-benefits beyond pollution reduction and environmental integrity, such as improving human health and well being.

Technological solutions such as improving recycling processes and developing nonharmful material degradability are needed, along with replacement by natural biodegradable materials.

With growing evidence of environmental consequences and potential threats to human health, we must consider industry's level of responsibility for the impacts of plastics.

Immediate actions are needed and possible.

Given these findings and the material shared at **MICRO** 2016 (see Appendix I for full program), we declare:

There is profound concern on the part of the scientific community about microplastics, which are clearly impacting the biosphere.

In recognition of the fact that microplastics continue accumulating and increasing, we must address the questions raised through the research presented here and continue expanding our knowledge horizons. This requires collaboration and cooperation, at all scales, from local to global, spanning sectors and disciplines, to improve knowledge, education and outreach efforts. This should not delay action.

With this declaration, we recognize our responsibility as individuals to change our behaviors related to plastic production and consumption, and to inform others of the social, economic, and environmental implications highlighted by the research shared at **MICRO** 2016.

As representatives of the scientific community, we urgently call upon society, the private sector and policymakers to move from knowledge to action.

This declaration marks the first milestone of the **Road to MICRO 2018** collaborative process.

PART IV

Breaking Down the Plastic Age

The following text is the academic version of the Lanzarote Declaration.

Breaking Down the Plastic Age

J. Baztan[1], M. Bergmann[2], A. Booth[3], E. Broglio[4,30], A. Carrasco[5], O. Chouinard[6], M. Clüsener-Godt[7], M. Cordier[1], A. Cozar[8], L. Devrieses[9], H. Enevoldsen[10], R. Ernsteins[11], M. Ferreira-da-Costa[12], M-C. Fossi[13], J. Gago[14], F. Galgani[15], J. Garrabou[4], G. Gerdts[2], M. Gomez[16], A. Gómez-Parra[8], L. Gutow[2], A. Herrera[16], C. Herring[17], T. Huck[18], A. Huvet[19], J-A. Ivar do Sul[20], B. Jorgensen[21], A. Krzan[22], F. Lagarde[23], A. Liria[16], A. Lusher[24], A. Miguelez[5], T. Packard[16], S. Pahl[25], I. Paul-Pont[18], D. Peeters[19], J. Robbens[26], A-C. Ruiz-Fernández[27], J. Runge[28], A. Sánchez-Arcilla[29], P. Soudant[18], C. Surette[6], R.C. Thompson[25], L. Valdés[7], J-P. Vanderlinden[1] and N. Wallace[17]

[1]Université de Versailles SQY, OVSQ, CEARC, Guyancourt, France [2]Alfred Wegener Institute for Polar and Marine Research (AWI), Bremerhaven, Germany [3]SINTEF Materials and Chemistry, Trondheim, Norway [4]Institute of Marine Sciences-CSIC, Barcelona, Spain [5]Cabildo de Lanzarote, Arrecife, Spain [6]Université de Moncton, Moncton, NB, Canada [7]UNESCO, Paris, France [8]Campus Universitario de Puerto Real, Cádiz, Spain [9]Animal Science Unit – Aquatic Environment and Quality, Oostende, Belgium [10]University of Copenhagen, Copenhagen, Denmark [11]University of Latvia, Rīga, Latvia [12]Universidad Federal de Pernambuco, Recife, Brazil [13]University of Siena, Siena, Italy [14]Instituto Español de Oceanografía, Vigo, Spain [15]IFREMER/Laboratoire Environnement Ressources PAC/Corse Imm Agostini, Bastia, France [16]Universidad de Las Palmas de Gran Canaria (ULPGC), Las Palmas, Spain [17]NOAA Office of Response and Restoration, Silver Spring, MD, United States [18]UBO-CNRS-LPO, Brest, France [19]IFREMER/Laboratoire Physiologie des Invertébrés, Plouzané, France [20]Universidade Federal do Rio Grande, Rio Grande, Brazil [21]Cornell University, Ithaca, NY, United States [22]National Institute of Chemistry, Ljubljana, Slovenia [23]Institut des Molécules et Matériaux de Mans (IMMM), Le Mans, France [24]National University of Ireland- Galway, Galway, Ireland [25]Plymouth University, Plymouth, United Kingdom [26]Institute for Agricultural and Fisheries Research, Merelbeke, Belgium [27]Instituto de Ciencias del Mar y Limnología, UNAM, Mazatlan, Mexico [28]Gulf of Maine Research Institute, Portland, ME, United States [29]Universitat Politècnica de Catalunya, Barcelona, Spain [30]Marine Sciences For Society.

Plastic materials have only been mass-produced for roughly 60 years, but nearly all aspects of our daily lives now involve plastics. Plastics are versatile, light, durable, and inexpensive materials that can be shaped to almost any form imaginable. While these are valuable characteristics, the exponential increase in the production and use of "disposable" plastic items has environmental consequences on a global scale. Plastics have been found in the air, soil, fresh water, seawater, deep-sea sediments, and sea ice. They are recognized as new habitat for organisms. Concerns about plastics are mounting due to their effects on organisms, economies, and human wellbeing. Conventional plastics are persistent materials that accumulate in the environment since they cannot be easily mineralized. Over time, larger items fragment in to smaller pieces. These small fragments currently outnumber larger, more visible pieces of plastic debris in the environment. This is particularly problematic because as it decreases in size, plastic pollution becomes increasingly difficult to

MICRO 2016. Fate and Impact of Microplastics in Marine Ecosystems.

remove from the environment and is more accessible for organisms to ingest, with largely unknown impacts.

Microplastics are generally defined as pieces of plastic debris <5 mm in size. They enter the environment either as primary or secondary microplastics. Primary microplastics—plastics manufactured in the microplastic size range—include: industrial abrasives, exfoliants, cosmetics, and preproduction plastic pellets. Secondary microplastics result from larger plastic items breaking down into smaller and smaller fragments.

MICRO 2016, the first international conference on microplastics, was held in the Biosphere Reserve of Lanzarote, Canary Islands, Spain. The MICRO 2016 Organizing Board and Scientific Committee brought together representatives from 30 research institutions and organizations. Following seminars in Plouzané, France in 2014 and Piran, Slovenia in 2015, MICRO 2016 summarized current knowledge and identified gaps and new questions. These were shared in the Lanzarote Declaration, released June 21, 2016 (see supplementary materials for full program and the Lanzarote Declaration).

The Lanzarote Declaration stems from previous regional, national, and international efforts such as: The London Convention (1972); the Barcelona Convention (1976); the MARPOL Convention (1978); the East Asian Seas Action Plan (1981); the Abidjan Convention (1984); the Cartagena Convention (1986); Bâle Convention (1989); the OSPAR Convention (1992/1998/2002/2005/2006/2007); the Northwest Pacific Action Plan (1994); the Nairobi Convention (1996); EU Water Framework Directive (2000); the Teheran Convention (2003); EU Marine Strategy Framework Directive (2008); the Honolulu Commitment (2011); the Manila Declaration (2012); the Mediterranean Regional Plan on Marine Litter (2014); and the G7 Leaders' Declaration (2015).

The 46 members of the Scientific Committee issued the Lanzarote Declaration on behalf of 632 researchers represented at the MICRO 2016 conference. The Declaration includes the following highlights from the scientific and technical research findings presented during MICRO 2016:

- Microplastics are found everywhere that has been investigated in the world, including the most remote parts of the earth (e.g., Compa et al., Russell et al., Frias et al., Martin et al., Palatinus et al., Hazimah et al., Nel et al., Naidoo et al., Reisser et al.,

Schoeneich-Argent et al., Buceta et al., Hajbane et al., van der Hal et al., Aytan et al., Chubarenko et al., Chouinard et al.). Though less studied, they are also present in fresh water, the atmosphere and terrestrial environments.

- Aquatic biota feeding grounds overlap with waters containing high levels of microplastic pollution. For example, this has been demonstrated for fin whales (*Balaenoptera physalus*) in the northwestern Mediterranean Sea (Fossi et al.) and inferred for marine mammals stranded in Ireland (Lusher et al.), sea turtles in Northern Cyprus (Duncan et al.), sea turtles in the Canary Current (Ostiategui et al.), and seabirds (e.g., Hidalgo-Ruz et al., Kühn et al.).
- Microplastics are ingested by many species, and the risk of transfer to humans has been shown for some commercial species such as fish (e.g., Pattersen et al., Scholz-Böttcher et al., Budimir et al.), mollusks, and crustaceans. Seaweed is also a demonstrated vector for microplastics (Gutow et al.). The potential consequences of microplastics in seafood is a growing concern and emerging research topic.
- Recent studies have confirmed that macro- and microplastics can be degraded into nanosized plastic particles, tens of microns in size or even smaller, with widely unknown impacts. Nanoplastic pollution will potentially be of significant importance in years to come.
- Very small microplastics or nanoplastics carrying chemicals or additives may be able to cross cell membranes and may enhance chemical bioavailability and stimulate toxic effects (e.g., Booth et al., Mintenig et al.).
- Runoff has been robustly demonstrated as a significant microplastic pathway, particularly road runoff in populated areas (e.g., Horton et al.). Several studies confirm sewage sludge as a pathway of microplastic pollution, highlighting the need for further studies and actions focused on sewage treatment plants and the urban water cycle (e.g., Maes et al., Mahon et al., Dris et al., Murphy et al., Crawford et al., Quinn et al.).
- While microorganisms will colonize any surface in seawater, new results confirm plastisphere microbial communities and provide site-specific insight into these communities and their successional changes over time (e.g., Gundry et al.).
- Modeling is clearly a fundamental and complementary tool for identifying microplastic sources, distribution paths, and potential sinks (e.g., van Sebille et al., Sanchez-Arcilla et al., Palatinus et al.).
- In order to integrate data across various studies and ongoing projects, we must: (1) standardize the identification and measurements of

microplastics (e.g., Gerdts et al., Fischer et al.); and (2) explicitly describe the techniques and methods currently used in ongoing, nonstandardized studies.

- Citizen science contributes to microplastics sampling and monitoring (e.g., Barrows, Boertien et al., Galgani et al.). Outreach and education efforts to raise awareness about microplastics in marine environments and increase ocean and plastic literacy help connect the general public with the issue of microplastics (e.g., Clusener-Godt et al., Jimenez et al., Ruckstuhl et al., Silva et al.).

- Perceptions and representations can be changed through science communication (e.g., Pahl et al., Jorgensen).

- Professional and citizen science efforts need to be standardized and harmonized, keeping in mind the importance of documenting the co-benefits of citizen science and the need for standardized databases and interfaces to share the results of citizen science work.

- Working to prevent and mitigate macro- and microplastic pollution can provide co-benefits beyond pollution reduction and environmental integrity, such as improving human health and wellbeing (e.g., Wyles et al.).

- Technological solutions such as improving recycling processes and developing nonharmful material degradability are needed, along with cautious exploration of natural biodegradable materials.

- With growing evidence of environmental consequences and potential threats to human health, we must consider the levels of responsibility governments and industries have for the impacts of microplastics.

- There is a need to maintain and improve the link between ongoing research and policy efforts at national and international levels, such as the EU Marine Strategy Framework Directive, OSPAR, NOWPAP, MEDPOL, etc.

- Immediate actions are needed and possible.

These findings and other material shared at MICRO 2016 (see supplementary materials for full program and references) demonstrate the profound concern of the scientific community regarding microplastics, which are clearly impacting the biosphere.

As microplastics continue accumulating in the environment, the scientific community must join forces to expand our knowledge horizons. Doing so requires collaboration and cooperation, at all scales, from local to global, spanning sectors and disciplines, to improve knowledge, education, and outreach efforts. This should not delay action.

As representatives of the scientific community, we urgently call upon the general public, policymakers at all levels, mediators, the media, educators, NGOs, entrepreneurs, and the private sector to move from assessment and knowledge to immediate action.

With the Lanzarote Declaration, we recognize our responsibility as individuals to change our behaviors related to plastic production and consumption, and to inform others of the social, cultural, economic, and environmental implications highlighted at MICRO 2016. The Lanzarote Declaration also serves as a milestone in working constructively as a research community to help stem the rising tide of plastics in the environment.

The MICRO community is engaging the challenge to work collaboratively. One reflection of this commitment will take place every two years in the form of an international conference, a forum to share available knowledge, fill in gaps, and establish new commitments for implementing solutions.

Microplastics in Lanzarote, Famara Beach, May 2016.

PART V

Where Next? The Road to Micro 2018 and Beyond

Where Next? The Road to Micro 2018 and Beyond

R.C. Thompson
Plymouth University, Plymouth, United Kingdom

Research on the sources, fate, and impacts of microplastic has increased exponentially over the last decade. As with any emerging discipline there are more questions than answers. Microplastics have been reported on shorelines, in the water column and in biota from the shallows to the deep sea, from the poles to the equator. There is also growing evidence of microplastics in freshwater habitats and elsewhere. However our ability to describe temporal and spatial patterns is limited and our understanding about the potential for microplastics to cause harmful effects in the natural environment is minimal. So what are the priorities for future research, what would we like to know by 2018, as well as over longer time scales?

In terms of distribution, the last decade has seen many pioneering studies demonstrating the presence of microplastics, and it is now clear that microplastics are widely distributed in the environment. We now need to move beyond these ground breaking descriptive studies to a more consistent, rigorous, and quantitative science. There is a need for standardization of protocols for the collection, separation, and identification of microplastics. Even before that, there is a need for consistency on the definition of microplastics. GESAMP took the view that microplastics are particles ranging between 5 mm and 1 nm. This is perfectly acceptable as working definition, but then we need to be clear that it is extremely challenging to isolate and identify particles less than 20 µm and although many suspect it, the presence of nanoplastic particles is yet to be confirmed in environmental samples. So there is a need to accept that any standardization of methods, especially for routine monitoring, will only address part of the picture. This is problematic because from a policy perspective it is essential to monitor in order to know whether levels of contamination are changing over time or between locations. However, the rationale for contaminant monitoring is usually to record the concentration or abundance of something that is considered potentially harmful. As yet we do not know which size fraction, which polymers and/or which co-contaminants present

MICRO 2016. Fate and Impact of Microplastics in Marine Ecosystems.

the greatest or lowest risks. Nor do we know which species or populations, if any, might be most affected. So while it is important to move toward intercomparability and standardization, it must be recognized that as yet the science is still exploratory and question driven. In my view it would be a mistake to define microplastics, microplastic monitoring and the scope of our interest simply on the basis of operational criteria such as the material captured by a particular net that might be widely in use for plankton sampling. We can use such nets and it would be good to standardize our approaches for comparability, but we need to acknowledge that we may later find that the particular form (size, shape, polymer, etc.) of microplastic that is of environmental concern is not well captured by such an approach. So there is a need for harmonization, but alongside this there is a need for further research to give clarity on the potential environmental issues that lead to the requirement for monitoring in the first place.

Turning to potential environmental impacts; experiments have clearly demonstrated the potential for plastics to transfer chemicals to organisms upon ingestion. However, there is uncertainty as to whether the relative importance of this pathway is sufficient to cause any additional toxicological harm, above that from the uptake of chemicals by ingestion of food or directly from the surrounding seawater. There is evidence of the potential for harm associated with the ingestion of microplastics alone, in the absence of any chemical transfer. It is important to recognize that with such a wide range of polymer types, sizes, and shapes together with various co-contaminants, many permutations remain untested and there is the potential for synergistic effects of chemical and particle toxicity. There has been criticism about the relevance of some toxicological studies to concentration of microplastic measured in the environment. It is clear that most laboratory experiments have used concentrations higher than those documented in the environment, however I am not sure this is either surprising or problematic. Much of the data we have on environmental concentrations is likely an underestimate of the true concentration because of the methodological limitations outlined. From a policy perspective it is important to understand the thresholds above which a substance might be harmful and then to set these into context with environmental concentrations as best we can. Our understanding of harm is very limited and, in my view, a logical approach is to start with experiments at high concentrations and then to progressively work down to lower concentrations. The key thing is to make sure the work we do is accurately put into perspective by clearly stating how the experiments relate to concentrations typically encountered in the environment. Notwithstanding this point there is a clear need to understand long-term chronic effects of low dose exposures and little has been done to address this.

So there is much to be done in order to fully understand the accumulation and potential impacts of microplastic in the environment. Having said that, in my view, any lack of information about microplastics is no reason not to take action to reduce the accumulation of plastic in the environment. Some action has been taken to reduce unnecessary emissions of microplastic directly to the environment, for example legislation to prevent the use of plastic particles in cosmetics. Further action may be needed to address other direct inputs of microplastic. However it is also important to take action to reduce the inputs of large items of plastic since these will fragment into the microplastics we will find in the decades to come. It is already clear that large items present a hazard to wildlife, that they have negative economic consequences and have the potential to damage human wellbeing. Plastics bring many societal benefits, and these can, in theory at least, be realized without the need for emissions of end-of-life plastics to the environment. Furthermore, if designed appropriately plastics are inherently recyclable. Since a substantial quantity of oil and gas are used to make plastics and the main application is in single use packaging, much of which enters the environment as litter, there is a clear need to prevent emissions and act in a more sustainable manner by moving toward a more circular economy—capturing end-of-life plastic so as to prevent it becoming waste or litter and at the same time reducing our reliance on nonrenewable resources. Research on the economic and social drivers and pathways that result in the accumulation of waste and litter in general are therefore needed as part of the wider picture moving towards solutions. Such solutions can only be reached by linking academia, industry, and policy so as to ensure potential actions are evaluated from a range of perspectives. There is already clear evidence that not doing so may simply generate unwanted consequences, for example "degradable" plastics designed with a view to reducing environmental impacts, but which in reality merely fragment into millions of pieces of microplastic.

In my view the accumulation of plastic and microplastic is an avoidable problem. To maximize the benefits that plastic can bring in reducing our footprint on the planet, it is therefore important that we act to reduce waste and litter. It is clear that the public are concerned about issues relating to the accumulation of plastic in the marine environment. If only we could harness this enthusiasm and focus it back at the problems which lie on the land. These are some of the challenges we need to work toward as we move to Micro 2018.

LIST OF CONTRIBUTORS

B. Abbas
Delft University of Technology (TU Delft), BC Delft, Netherlands

A. Abreu
UNESCO, Paris, France

R. Adams
Plymouth University, Plymouth, United Kingdom

M. Adolfsson-Erici
Stockholm University, Stockholm, Sweden

A. Afonso
European Food Safety Authority (EFSA), Parma, Italy

C. Afonso-Olivares
Universidad de Las Palmas de Gran Canaria, Las Palmas de Gran Canaria, Spain

E. Agirbas
Recep Tayyip Erdogan University (RTEU), Rize, Turkey

J.M. Aguiló
Agència Balear de l'Aigua i la Qualitat Ambiental (ABAQUA), Palma de Mallorca, Spain

L. Airoldi
University of Bologna, Ravenna, Italy

H. Aksoy
Sakarya University, Sakarya, Turkey

M. Albentosa
Spanish Institute of Oceanography, Murcia, Spain

L. Alcaro
ISPRA, Italian National Institute for Environmental Protection and Research, Roma, Italy

S. Aliani
CNR-ISMAR, La Spezia, Italy

I. Al-Maslamani
Qatar University, Doha, Qatar

C. Alomar
Instituto Español de Oceanografía, Palma de Mallorca, Spain; Instituto Español de Oceanografía, Madrid, Spain

D. Altin
Biotrix, Trondheim, Norway

E. Álvarez
Ente Público Puertos del Estado, Madrid, Spain

L.A. Amaral-Zettler
Josephine Bay Paul Center for Comparative Molecular Biology and Evolution, Woods Hole, MA, United States; Brown University, Providence, RI, United States

E. Amato
ISPRA, Italian National Institute for Environmental Protection and Research, Roma, Italy

A. Anderson
Plymouth University, Plymouth, United Kingdom

A.L. Andrady
North Carolina State University, Raleigh, NC, United States

G. Andrius
Center for Physical Science and Technology (FTMC), Vilnius, Lithuania

D. Angel
University of Haifa, Haifa, Israel

F. Ariese
VU Amsterdam, Amsterdam, The Netherlands

H.P. Arp
Norwegian Geotechnical Institute (NGI), Oslo, Norway

M. Asensio
Universidad de Las Palmas de Gran Canaria, Las Palmas de Gran Canaria, Spain

K. Assidqi
Bogor Agricultural University, Bogor, Indonesia

C.G. Avio
Università Politecnica delle Marche, Ancona, Italy

U. Aytan
Recep Tayyip Erdogan University (RTEU), Rize, Turkey

T. Bahri
Food and Agricultural Organization of the United Nations (FAO), Rome, Italy

M. Baini
University of Siena, Siena, Italy

A. Bakir
Plymouth University, Plymouth, United Kingdom

H. Ball
Moores University, Liverpool, United Kingdom

C. Baranyi
University of Vienna, Vienna, Austria

L.G.A. Barboza
University of Porto, Porto, Portugal; CIIMAR/CIIMAR-LA, Interdisciplinary Centre of Marine and Environmental Research, Porto, Portugal

U. Barg
Food and Agricultural Organization of the United Nations (FAO), Rome, Italy

L. Bargelloni
Università di Padova, Padova, Italy

H. Barras
Heriot-Watt University (HWU), Edinburgh, Scotland

C. Barrera
Oceanic Platform of the Canary Islands (PLOCAN), Telde, Spain

P. Barria
Universidade de Coimbra, Coimbra, Portugal

A. Barrows
Adventure Scientists, Bozeman, MT, United States; College of the Atlantic (COA), Bar Harbor, ME, United States

A. Barth
Stockholm University, Stockholm, Sweden

A. Batel
University of Heidelberg, Heidelberg, Germany

J. Baztan
Université de Versailles Saint-Quentin-en-Yvelines, Guyancourt, France; Marine Sciences For Society

P. Baztan
Marine Sciences For Society; Universitat Autònoma de Barcelona (UAB), Barcelona, Spain

R. Beiras
University of Vigo, Pontevedra, Spain

M. Benedetti
Università Politecnica delle Marche, Ancona, Italy

A.A. Berber
Sakarya University, Sakarya, Turkey

N. Berber
Sakarya University, Sakarya, Turkey

M. Bergmann
Alfred Wegener Institute, Helmholtz Centre for Polar and Marine Research (AWI), Bremerhaven, Germany; Alfred Wegener Institute, Helmholtz Centre for Polar and Marine Research, Helgoland, Germany

M. Berlino
Universita Politecnica delle Marche, Ancona, Italy

S. Berrow
Irish Whale and Dolphin Group, Kilrush, Ireland

F. Bessa
Universidade de Coimbra, Coimbra, Portugal

E. Besseling
Institute for Marine Resources & Ecosystem Studies (IMARES) Wageningen UR, IJmuiden, Netherlands

B. Beyer
Alfred Wegener Institute, Helmholtz Centre for Polar and Marine Research (AWI), Bremerhaven, Germany

M. Binaglia
European Food Safety Authority (EFSA), Parma, Italy

T. Bizjak
University of Eastern Finland, Kuopio, Finland

K.A. Bjorndal
University of Florida, Gainesville, FL, United States

R. Blust
University of Antwerp, Antwerpen, Belgium

M. Boertien
Ocean Conservation

A.B. Bolten
University of Florida, Gainesville, FL, United States

A.M. Booth
SINTEF Materials and Chemistry, Trondheim, Norway

B. Bounoua
University of Paris-Est, Créteil, France

P. Bourseau
Université Bretagne Sud, Lorient, France; Université de Nantes, Saint-Nazaire, France

N. Brahimi
Rennes School of Business, Rennes, France

M. Bramini
IIT, Italian Institute of Technology, Genova, Italy

N. Brennholt
German Institute for Hydrology, Koblenz, Germany

E. Breuninger
German Institute for Hydrology, Koblenz, Germany

J. Bried
MARE – Marine and Environmental Sciences Centre, Horta, Portugal; IMAR – Instituto do Mar, Horta, Portugal

A. Broderick
University of Exeter, Penryn, United Kingdom

E. Broglio
Marine Sciences For Society; Institute of Marine Sciences-CSIC, Barcelona, Spain

M.A. Browne
University of California, Santa Barbara, CA United States; University of New South Wales, Sydney, NSW, Australia

S. Bruzaud
Université Bretagne Sud, Lorient, France

J. Buceta
Centro de Estudios de Puertos y Costas, Madrid, Spain

S. Buchinger
German Institute for Hydrology, Koblenz, Germany

S. Budimir
Finnish Environment Institute, Helsinki, Finland

H. Budzin-ski
Université de Bordeaux, Bordeaux, France

E. Butter
Ocean Conservation

J. Cachot
University of Bordeaux, Talence, France

M. Caetano
Instituto Português do Mar e da Atmosfera, Lisbon, Portugal

A. Callaghan
University of Reading, Reading, United Kingdom

A. Camedda
Institute for Coastal Marine Environment and National Research Council (IAMC-CNR),
Oristano, Italy

S. Canella
Autoridad Portuaria de Las Palmas, Las Palmas de Gran Canaria, Spain

L. Cardelli
Università Politecnica delle Marche, Ancona, Italy

S. Carpentieri
LEGAMBIENTE, Roma, Italy

A. Carrasco
Observatorio Reserva de Biosfera, Arrecife, Spain; Cabildo de Lanzarote, Arrecife, Spain

R. Carriço
Universidade dos Açores, Horta, Portugal; MARE − Marine and Environmental Sciences
Centre, Horta, Portugal; IMAR − Instituto do Mar, Horta, Portugal

A. Caruso
Université du Maine, Le Mans, France

A.-L. Cassone
Institut Français de Recherche pour l'Exploitation de la Mer (IFREMER), Plouzané, France

A. Castillo
Qatar University, Doha, Qatar

R.O. Castro
Universidade Federal Fluminense, Niterói, Brazil

A.I. Catarino
Heriot-Watt University (HWU), Edinburgh, Scotland

P.W. Cazenave
Plymouth Marine Laboratory, Plymouth, United Kingdom

İ. Çelik
Sakarya University, Sakarya, Turkey

P. Cerralbo
Universitat Politècnica de Catalunya, Barcelona, Spain

G. César
Université Bretagne Sud, Lorient, France

O. Chouinard
Marine Sciences For Society; Université de Moncton, Moncton, NB, Canada

I. Chubarenko
P.P. Shirshov Institute of Oceanology of Russian Academy of Sciences, Kaliningrad, Russia

I.P. Chubarenko
Atlantic Branch of P.P. Shirshov Institute of Oceanology, Russian Academy of Sciences, Kaliningrad, Russia

A.M. Cicero
ISPRA, Italian National Institute for Environmental Protection and Research, Roma, Italy

G. Clarindo
University of Las Palmas de Gran Canaria (ULPGC), Las Palmas de Gran Canaria, Spain

B. Clarke
RMIT University, Bundoora, VIC Australia

C. Clérandeau
University of Bordeaux, Talence, France

M. Clüsener-Godt
UNESCO, Paris, France

M. Codina-García
Universitat de Barcelona, Barcelona, Spain

M. Cole
Plymouth Marine Laboratory, Plymouth, United Kingdom; University of Exeter, Exeter, United Kingdom

F. Collard
University of Liege, Liège, Belgium

A. Collignon
University of Liege, Liège, Belgium

T. Collins
Sea Education Association, Woods Hole, MA, United States

M. Compa
Instituto Español de Oceanografía, Palma de Mallorca, Spain; Instituto Español de Oceanografía, Madrid, Spain

P. Conan
Université Pierre et Marie Curie (UPMC), Paris, France

M. Cordier
Université de Versailles Saint-Quentin-en-Yvelines, Guyancourt, France

W. Courtene-Jones
Scottish Association for Marine Science, Oban, Scotland

X. Cousin
IFREMER, Plouzané, France

P. Covelo
Coordinadora para o Estudio dos Mamíferos Mariños, CEMMA, Gondomar, Spain

A. Cózar
Universidad de Cádiz, Cádiz, Spain

E. Crichton
Vancouver Aquarium Marine Science Centre, Vancouver, BC, Canada

O. Crispi
Université Pierre et Marie Curie (UPMC), Paris, France

M. Cronin
Coastal & Marine Research Centre (CMRC), UCC, Cork, Ireland

P.L. Croot
National University of Ireland Galway, Galway, Ireland

M.J. Cruz
OMA – Observatório do Mar dos Açores, Horta, Portugal

G. d'Errico
Università Politecnica delle Marche, Ancona, Italy

C. Dâmaso
OMA   Observatório do Mar dos Açores, Horta, Portugal

K. Das
University of Liege, Liège, Belgium

L.F. de Alencastro
Central Environmental Laboratory, EPFL, Lausanne, Switzerland; Swiss Federal Institute of Technology, Lausanne, Switzerland

F.V. de Araujo
Universidade Federal Fluminense, Niterói, Brazil; Universidade do Estado do Rio de Janeiro, Rio de Janeiro, Brazil

J.F. de Boer
VU Amsterdam, Amsterdam, The Netherlands

G.A. de Lucia
Institute for Coastal Marine Environment and National Research Council (IAMC-CNR), Oristano, Italy

P. Debeljak
The Ocean Cleanup (TOC), ES Delft, Netherlands

A. Dehaut
Agence nationale de sécurité sanitaire de l'alimentation, de l'environnement et du travail (Anses), Boulogne sur Mer, France

S. Deudero
Instituto Español de Oceanografía, Palma de Mallorca, Spain; Instituto Español de Oceanografía, Madrid, Spain

L. Devrieses
Animal Science Unit – Aquatic Environment and Quality, Oostende, Belgium

S. Di Vito
LEGAMBIENTE, Roma, Italy

A. Díaz
Vancouver Aquarium Marine Science Centre, Vancouver, BC, Canada; Universidad de Cádiz, Cádiz, Spain

J. Donohue
Sea Education Association, Woods Hole, MA, United States

P. Doumenq
Aix-Marseille Université, Aix-en-Provence, France

T.K. Doyle
National University of Ireland Galway, Galway, Ireland

R. Dris
University of Paris-Est, Créteil, France

J.-N. Druon
Institute for the Protection and Security of the Citizen (IPSC), Ispra, Italy

C.M. Duarte
King Abdullah University of Science and Technology, Thuwal, Kingdom of Saudi Arabia

G. Duflos
Agence nationale de sécurité sanitaire de l'alimentation, de l'environnement et du travail (Anses), Boulogne sur Mer, France

E. Duncan
University of Exeter, Penryn, United Kingdom; Plymouth Marine Laboratory, Plymouth, United Kingdom

C. Dussud
Université Pierre et Marie Curie (UPMC), Paris, France

A. Eckerlebe
Alfred Wegener Institute Helmholtz Centre for Polar and Marine Research (AWI), Bremerhaven, Germany

M. Egelkraut-Holtus
Shimadzu Europa GmbH, Duisburg, Germany

D.P. Eidsvoll
Norwegian Institute for Water Research (NIVA), Oslo, Norway.

C. Ek
Stockholm University, Stockholm, Sweden

S. Elena
The A. O. Kovalevsky Institute of Marine Biological Research of RAS, Sevastopol, Russian Federation

A. Elineau
Université Pierre et Marie Curie (UPMC), Paris, France

H. Enevoldsen
University of Copenhagen, Copenhagen, Denmark

G. Eppe
University of Liege, Liège, Belgium

M. Eriksen
Five Gyres Institute, Los Angeles, CA, United States

R. Ernsteins
University of Latvia, Rīga, Latvia

M. Espino
Universitat Politècnica de Catalunya, Barcelona, Spain

N. Estévez-Calvar
National Council of Researches (CNR), Genova, Italy

C. Ewins
University of the West of Scotland, Paisley, Scotland

P. Fabre
Université de Montpellier, Montpellier, France

M. Faimali
National Council of Researches (CNR), Genova, Italy

D. Fattorini
Università Politecnica delle Marche, Ancona, Italy

F. Faure
Central Environmental Laboratory, EPFL, Lausanne, Switzerland; Swiss Federal Institute of Technology, Lausanne, Switzerland

D. Ferrando
University of Genova, Genova, Italy

J.C. Ferreira
New University of Lisbon, Lisbon, Portugal

M. Ferreira-da-Costa
Universidad Federal de Pernambuco, Recife, Brazil

E. Fileman
Plymouth Marine Laboratory, Plymouth, United Kingdom

M. Fischer
University of Oldenburg, Oldenburg, Germany

A.B. Fortunato
National Laboratory for Civil Engineering, Lisbon, Portugal

M.C. Fossi
University of Siena, Siena, Italy

V. Foulon
Institut Universitaire Européen de la Mer (IUEM), Plouzané, France

A. Frank
Instituto Español de Oceanografía, Palma de Mallorca, Spain

M. Frenzel
SINTEF Materials and Chemistry, Trondheim, Norway

L. Frère
Institut Français de Recherche pour l'Exploitation de la Mer (IFREMER), Plouzané, France; Institut Universitaire Européen de la Mer (IUEM), Plouzané, France

J.P.G.L. Frias
Universidade dos Açores, Horta, Portugal; MARE – Marine and Environmental Sciences Centre, Horta, Portugal; IMAR – Instituto do Mar, Horta, Portugal

H. Frick
University of Copenhagen, Frederiksberg, Denmark

P.W. Froneman
Rhodes University, Grahamstown, South Africa

V.M. Gabet
Instituto Tecnológico de Canarias, Las Palmas, Spain

G.W. Gabrielsen
Norwegian Polar Institute, Tromsø, Norway

J. Gago
Instituto Español de Oceanografía (IEO), Vigo, Spain

T. Gajst
University of Eastern Finland, Kuopio, Finland

F. Galgani
Institut Français de Recherche pour l'Exploitation de la Mer (IFREMER), Bastia, France;
University of Siena, Siena, Italy

M. Gallinari
Institut Universitaire Européen de la Mer (IUEM), Plouzané, France

T.S. Galloway
University of Exeter, Exeter, United Kingdom

E.G. Gamarro
Food and Agricultural Organization of the United Nations (FAO), Rome, Italy

C. Gambardella
National Council of Researches (CNR), Genova, Italy

F. Garaventa
National Council of Researches (CNR), Genova, Italy

S. Garcia
DRAM-Direção Regional dos Assuntos do Mar, Horta, Portugal

J. Garrabou
Marine Sciences For Society; Institute of Marine Sciences-CSIC, Barcelona, Spain

P. Garrido
Universidad de Las Palmas de Gran Canaria, Las Palmas de Gran Canaria, Spain

S.F. Gary
Scottish Association for Marine Science, Oban, Scotland

J. Gasperi
University of Paris-Est, Créteil, France

W. Gaze
University of Exeter, Exeter, United Kingdom

T. Geertz
Plastic Change, Hellerup, Denmark

M.D. Gelado-Caballero
University of Las Palmas de Gran Canaria (ULPGC), Las Palmas de Gran Canaria, Spain

M. George
Université de Montpellier, Montpellier, France

J. Gercken
Institute for Applied Ecology (IfAÖ), Neu Broderstorf, Germany

G. Gerdts
Alfred Wegener Institute, Helmholtz Centre for Polar and Marine Research (AWI), Bremerhaven, Germany; Alfred Wegener Institute, Helmholtz Centre for Polar and Marine Research (AWI), Helgoland, Germany

J.-F. Ghiglione
Université Pierre et Marie Curie (UPMC), Paris, France

E. Gies
Vancouver Aquarium Marine Science Centre, Vancouver, BC, Canada

B. Gilbert
University of Liege, Liège, Belgium

L. Giménez
Bangor University, Bangor, United Kingdom

D. Glassom
University of KwaZulu-Natal (UKZN), Durban, South Africa

M. Glockzin
Leibniz Institute for Baltic Sea Research, Rostock, Germany

B. Godley
University of Exeter, Penryn, United Kingdom

K. Goede
Rap.ID Particle Systems GmbH, Berlin, Germany

A. Goksøyr
University of Bergen, Bergen, Norway

M. Gómez
Universidad de Las Palmas de Gran Canaria, Las Palmas de Gran Canaria, Spain

A. Gómez-Parra
Campus Universitario de Puerto Real, Cádiz, Spain

D. González-Marco
Universitat Politècnica de Catalunya, Barcelona, Spain

J. González-Solís
Universitat de Barcelona, Barcelona, Spain

S. Gorbi
Università Politecnica delle Marche, Ancona, Italy

E. Gorokhova
Stockholm University, Stockholm, Sweden

G. Gorsky
Université Pierre et Marie Curie (UPMC), Paris, France

M. Gosch
University College Cork, Cork, Ireland

J. Grose
Plymouth University, Plymouth, United Kingdom

G.M. Guebitz
Austrian Centre of Industrial Biotechnology, ACIB, Tulln, Austria; BOKU University of Natural Resources and Life Sciences, Vienna, Tulln, Austria

R. Guedes-Alonso
Universidad de Las Palmas de Gran Canaria, Las Palmas de Gran Canaria, Spain

B. Guijarro
Instituto Español de Oceanografía, Palma de Mallorca, Spain

L. Guilhermino
University of Porto, Porto, Portugal; CIIMAR/CIIMAR-LA, Interdisciplinary Centre of Marine and Environmental Research, Porto, Portugal

T. Gundry
RMIT University, Bundoora, VIC, Australia

L. Gutow
Alfred Wegener Institute, Helmholtz Centre for Polar and Marine Research (AWI), Bremerhaven, Germany; Alfred Wegener Institute, Helmholtz Centre for Polar and Marine Research (AWI), Helgoland, Germany

M. Haave
Uni Research, Bergen, Norway

M. Haeckel
GEOMAR Helmholtz-Zentrum für Ozeanforschung, Kiel, Germany

K. Haernvall
Austrian Centre of Industrial Biotechnology, ACIB, Tulln, Austria

S. Hajbane
The University of Western Australia (CEME), Crawley, WA, Australia

M. Hamann
James Cook University (JCU), Townsville, QLD, Australia

J. Hämer
Alfred Wegener Institute Helmholtz Centre for Polar and Marine Research (AWI), Bremerhaven, Germany

T. Hamm
GEOMAR, Helmholtz Center for Ocean Research in Kiel, Kiel, Germany

B.H. Hansen
SINTEF Materials and Chemistry, Trondheim, Norway

B.D. Hardesty
CSIRO Oceans and Atmosphere Flagship, Hobart, TAS, Australia

B. Harth
Alfred Wegener Institute Helmholtz Zentrum für Polar und Meeresforschung (AWI), Helgoland, Germany

S. Hartikainen
University of Eastern Finland (UEF), Kuopio, Finland

M. Hassellöv
University of Gothenburg, Gothenburg, Sweden

S. Hatzky
German Institute for Hydrology, Koblenz, Germany

M.G. Healy
National University of Ireland, Galway, Ireland

H. Hégaret
Institut Universitaire Européen de la Mer (IUEM), Plouzané, France

T.B. Henry
Heriot-Watt University (HWU), Edinburgh, Scotland

L. Hermabessiere
Agence nationale de sécurité sanitaire de l'alimentation, de l'environnement et du travail (Anses), Boulogne sur Mer, France

J.J. Hernández-Brito
Oceanic Platform of the Canary Islands (PLOCAN), Telde, Spain

A. Hernandez-Gonzalez
Instituto Español de Oceanografía, Vigo, Spain

G. Hernandez-Milian
University College Cork, Cork, Ireland

Ü. Herind
The Ocean Cleanup (TOC), ES Delft, Netherlands; University of Vienna, Vienna, Austria

A. Herrera
Universidad de Las Palmas de Gran Canaria (ULPGC), Las Palmas de Gran Canaria, Spain

C. Herring
NOAA Office of Response and Restoration, Silver Spring, MD, United States

D. Herzke
Norwegian Institute for Air Research, Tromsø, Norway

V. Hidalgo-Ruz
Universidad Católica del Norte, Coquimbo, Chile; Millennium Nucleus of Ecology and Sustainable Management of Oceanic Island (ESMOI), Coquimbo, Chile

C. Himber
Agence nationale de sécurité sanitaire de l'alimentation, de l'environnement et du travail (Anses), Boulogne sur Mer, France

M. Holland
Plymouth University, Plymouth, United Kingdom

N.-H. Hong
Korea Institute of Toxicology, Daejeon, Republic of Korea

A.A. Horton
Centre for Ecology and Hydrology, Wallingford, United Kingdom

P. Horvat
National Institute of Chemistry, Ljubljana, Slovenia

T. Huck
Marine Sciences For Society; UBO-CNRS-LPO, Brest, France

M. Huhn
Bogor Agricultural University, Bogor, Indonesia; GEOMAR Helmholtz Centre for Ocean Research Kiel, Kiel, Germany

A. Huvet
Institut Français de Recherche pour l'Exploitation de la Mer (IFREMER), Plouzané, France; Institut Universitaire Européen de la Mer (IUEM), Plouzané, France

M. Iglesias
Instituto Español de Oceanografía, Madrid, Spain

C. Igor
The A. O. Kovalevsky Institute of Marine Biological Research of RAS, Sevastopol, Russian Federation

I.A. Isachenko
Atlantic Branch of P.P. Shirshov Institute of Oceanology, Russian Academy of Sciences, Kaliningrad, Russia

J-A. Ivar do Sul
Universidade Federal do Rio Grande, Rio Grande, Brazil

A. Jahnke
Helmholtz Centre for Environmental Research – UFZ, Leipzig, Germany

B. Janis
University of Latvia, Riga, Latvia

K. Janis
University of Latvia, Riga, Latvia

U. Janis
University of Latvia, Riga, Latvia

A. Jemec
University of Ljubljana, Ljubljana, Slovenia

J.C. Jiménez
Oficina de la Reserva de la Biosfera de Lanzarote, Lanzarote (Canary Islands), Spain

H. Johnsen
SINTEF Materials and Chemistry, Trondheim, Norway

B. Jorgensen
Cornell University, Ithaca, NY, United States; Marine Sciences For Society

J.H. Jørgensen
NIRAS A/S, Allerød, Denmark

H. Jörundsdóttir
Matís ltd., Reykjavik, Iceland

Y.-J. Jung
Korea Institute of Toxicology, Daejeon, Republic of Korea

M. Kedzierski
Université Bretagne Sud, Lorient, France

S. Keiter
Orebro University, Örebro, Sweden

P. Kershaw
Food and Agricultural Organization of the United Nations (FAO), Rome, Italy

K. Kesy
Leibniz Institute for Baltic Sea Research, Rostock, Germany

F. Khan
Roskilde University, Roskilde, Denmark

L.I. Khatmullina
Atlantic Branch of P.P. Shirshov Institute of Oceanology, Russian Academy of Sciences, Kaliningrad, Russia

J. Kirby
Moores University, Liverpool, United Kingdom

K. Kiriakoulakis
Moores University, Liverpool, United Kingdom

R. Klein
Trier University, Trier, Germany

T. Klunderud
Uni Research, Bergen, Norway; University of Bergen, Bergen, Norway

C.M.H. Knudsen
Roskilde University, Roskilde, Denmark

T.B. Knudsen
Roskilde University, Roskilde, Denmark

C. Kochleus
German Institute for Hydrology, Koblenz, Germany

A.A. Koelmans
Wageningen University (WUR), Wageningen, The Netherlands

T. Kögel
The National Institute of Nutrition and Seafood Research (NIFES), Bergen, Norway

A. Koistinen
University of Eastern Finland (UEF), Kuopio, Finland

K. Kopke
University College Cork, Cork, Ireland

Š. Korez
Alfred Wegener Insitute Helmholtz Centre for Polar and Marine Research (AWI), Bremerhaven, Germany

N. Kowalski
Leibniz Institute for Baltic Sea Research, Rostock, Germany

B. Kreikemeyer
University of Rostock (UniR), Rostock, Germany

F. Kroon
Australian Institute of Marine Science, Townsville, QLD, Australia

T. Krumpen
Alfred Wegener Institute, Helmholtz Centre for Polar and Marine Research (AWI), Bremerhaven, Germany

A. Krzan
National Institute of Chemistry, Ljubljana, Slovenia

A. Kržan
National Institute of Chemistry, Ljubljana, Slovenia

M. Labrenz
Leibniz Institute for Baltic Sea Research, Rostock, Germany

C. Lacroix
CEDRE, Brest, France

L. Ladirat
Université Paul Sabatier, Toulouse, France

C. Laforsch
University of Bayreuth, Bayreuth, Germany

F. Lagarde
University of Maine, Le Mans, France

E. Lahive
Centre for Ecology and Hydrology, Wallingford, United Kingdom

C. Lambert
Institut Universitaire Européen de la Mer (IUEM), Plouzané, France; Institut Français de Recherche pour l'Exploitation de la Mer (IFREMER), Plouzané, France; Institut Universitaire Européen de la Mer (IUEM), Plouzané, France

C. Lapucci
LaMMA Consortium-CNR Ibimet, Sesto Fiorentino, Italy

G. Lattin
Algalita Marine Research and Education, Long Beach, CA, United States

K.L. Law
Sea Education Association, Woods Hole, MA, United States

F. Le Roux
Institut Français de Recherche pour l'Exploitation de la Mer (IFREMER), Plouzané, France; Sorbonne Universités, Roscoff, France

K. Le Souef
Vancouver Aquarium, Vancouver, BC, Canada

V. Le Tilly
Université Bretagne Sud, Lorient, France

L. Lebreton
Dumpark Data Science, Wellington, New Zealand

E. Leemans

M. Lehtiniemi
Finnish Environment Institute, Helsinki, Finland

M. Lenz
GEOMAR Helmholtz Centre for Ocean Research Kiel, Kiel, Germany

J. Leskinen
University of Eastern Finland, Kuopio, Finland

H. Leslie
VU University Amsterdam, Amsterdam, The Netherlands

H.A. Leslie
VU Amsterdam, Amsterdam, The Netherlands

C. Levasseur
Swiss Federal Institute of Technology, Lausanne, Switzerland

C. Lewis
University of Exeter, Exeter, United Kingdom

P. Licandro
Sir Alister Hardy Foundation for Ocean Science (SAHFOS), Plymouth, United Kingdom

K. Lind
Tallinn University of Technology, Tallinn, Estonia

P. Lindeque /
Plymouth Marine Laboratory, Plymouth, United Kingdom

P.K. Lindeque
Plymouth Marine Laboratory, Plymouth, United Kingdom

I. Lips
Tallinn University of Technology, Tallinn, Estonia

A. Liria
Marine Sciences For Society; Asociación para el desarrollo sostenible y biodiversidad (ADS Biodiversidad); Universidad de Las Palmas de Gran Canaria (ULPGC), Las Palmas, Spain

A. Liria-Loza
Asociación para el desarrollo sostenible y biodiversidad (ADS Biodiversidad)

O. Llinás
Plataforma Oceanica de Canarias (PLOCAN), Telde – Gran Canaria, Spain

S.A. Loiselle
University of Siena, Siena, Italy

M. Long
Institut Universitaire Européen de la Mer (IUEM), Plouzané, France

C. Lorenz
Alfred Wegener Institute Helmholtz Centre for Polar and Marine Research (AWI), Helgoland, Germany

S.M. Lorenzo
Universidade da Coruña (UDC)-Instituto, A Coruña, Spain

K. Loubar
Ecole des Mines de Nantes, Nantes, France

G. Luna-Jorquera
Universidad Católica del Norte, Coquimbo, Chile; Millennium Nucleus of Ecology and Sustainable Management of Oceanic Island (ESMOI), Coquimbo, Chile; Centro de Estudios Avanzados en Zonas Áridas (CEAZA), Coquimbo, Chile

A.L. Lusher
Galway-Mayo Institute of Technology, Galway, Ireland; National University of Ireland Galway, Galway, Ireland

V. Macchia
Heriot-Watt University (HWU), Edinburgh, Scotland

S. MacGabban
University College Cork, Cork, Ireland

K. Mackay
European Food Safety Authority (EFSA), Parma, Italy

M. MacLeod
Stockholm University, Stockholm, Sweden

T. Maes
CEFAS, Centre for Environment, Fisheries, Aquaculture and Science, Lowestoft, UK

E. Magaletti
ISPRA, Italian National Institute for Environmental Protection and Research, Roma, Italy

A. Maggiore
European Food Safety Authority (EFSA), Parma, Italy

K. Magnusson
IVL Swedish Environmental Research Institute, Göteborg, Sweden; IVL Swedish Environmental Research Institute, Stockholm, Sweden

A.M. Mahon
Galway-Mayo Institute of Technology, Galway, Ireland

L. Maiju
Finnish Environment Institute, Helsinki, Finland

P. Makorič
University of Nova Gorica, Nova Gorica, Slovenia

O. Mallow
Vienna University of Technology, Vienna, Austria

D. Marc
Parc naturel marin du Golfe du Lion, Port-Vendres, France

J. Marques
Universidade de Coimbra, Coimbra, Portugal

L. Marsili
University of Siena, Siena, Italy

E. Martí
Universidad de Cádiz, Puerto Real, Spain

M. Martignac
Université Paul Sabatier, Toulouse, France

J. Martin
National University of Ireland Galway, Galway, Ireland

I. Martínez
Universidad de Las Palmas de Gran Canaria, Las Palmas de Gran Canaria, Spain

J. Martínez
Universitat de Barcelona, Barcelona, Spain

M. Martinez-Gil
Division para la Proteccion del Mar., Madrid, Spain

H.R. Martins
MARE – Marine and Environmental Sciences Centre, Horta, Portugal; IMAR – Instituto do Mar, Horta, Portugal

M. Matiddi
ISPRA, Italian National Institute for Environmental Protection and Research, Roma, Italy

N. Maximenko
University of Hawai'i at Mānoa, Honolulu, HI, United States

R. Mazlum
Recep Tayyip Erdogan University (RTEU), Rize, Turkey

R. Mcadam
Imperial College London, London, United Kingdom

L. Mcknight
University of Las Palmas de Gran Canaria (ULPGC), Las Palmas de G.C., Spain; Institute of Environmental and Natural Resources Study (i-UNAT), ULPGC, Las Palmas de G.C., Spain

A.W. McNeal
Plymouth Marine Laboratory, Plymouth, United Kingdom

J. Measures
Heriot-Watt University (HWU), Edinburgh, Scotland

M.S. Mederos
Oficina de la Reserva de la Biosfera de Lanzarote, Lanzarote (Canary Islands), Spain

C. Mel
Université de Perpignan, Perpignan, France

J. Mendoza
Food and Agricultural Organization of the United Nations (FAO), Rome, Italy

M.S. Meyer
University of Oldenburg, Oldenburg, Germany

A. Miguelez
Observatorio Reserva de Biosfera, Arrecife, Spain; Cabildo de Lanzarote, Arrecife, Spain

M. Milan
Università di Padova, Padova, Italy

T. Militão
Universitat de Barcelona, Barcelona, Spain

R.Z. Miller
Rozalia Project, Granville, VT, United States

G. Mir
Oficina de la Reserva de la Biosfera de Lanzarote, Lanzarote (Canary Islands), Spain

D. Miranda-Urbina
Universidad Católica del Norte, Coquimbo, Chile; Millennium Nucleus of Ecology and Sustainable Management of Oceanic Island (ESMOI), Coquimbo, Chile

F. Misurale
National Council of Researches (CNR), Genova, Italy

S. Montesdeoca-Esponda
Universidad de Las Palmas de Gran Canaria, Las Palmas de Gran Canaria, Spain

J. Mora
Autoridad Portuaria de Santa Cruz de Tenerife, Santa Cruz de Tenerife, Spain

M.-V.-V. Morgan
Université de Perpignan, Perpignan, France; Parc naturel marin du Golfe du Lion, Port-Vendres, France

S. Morgana
National Council of Researches (CNR), Genova, Italy

B. Moriceau
Institut Universitaire Européen de la Mer (IUEM), Plouzané, France

B. Morin
University of Bordeaux, Talence, France

A. Morley
National University of Ireland Galway, Galway, Ireland

L. Morrison
National University of Ireland, Galway, Ireland

F. Murphy
University of the West of Scotland, Paisley, Scotland

T. Naidoo
University of KwaZulu-Natal (UKZN), Durban, South Africa

P. Näkki
Finnish Environment Institute, Helsinki, Finland

I.E. Napper
Plymouth University, Plymouth, United Kingdom

B.E. Narayanaswamy
Scottish Association for Marine Science, Oban, Scotland

R. Nash
Galway-Mayo Institute of Technology, Galway, Ireland

A. Negri
Australian Institute of Marine Science, Townsville, QLD, Australia

H.A. Nel
Rhodes University, Grahamstown, South Africa

M.S. Nerheim
University of Bergen (UIB), Bergen, Norway

I.L. Nerland
Norwegian Institute for Water Research (NIVA), Oslo, Norway.

J. Neto
Universidade de Coimbra, Coimbra, Portugal

V. Neves
MARE – Marine and Environmental Sciences Centre, Horta, Portugal; IMAR – Instituto do Mar, Horta, Portugal; University of the Azores, Horta, Portugal

H. Nies
Federal Maritime and Hydrographic Agency, Hamburg, Germany

M. Noel
Vancouver Aquarium Marine Science Centre, Vancouver, BC, Canada

N.H.M. Nor
National University of Singapore, Singapore, Singapore

F. Noren
IVL Swedish Environmental Research Institute, Fiskebäckskil, Sweden

B. O' Connell
Galway-Mayo Institute of Technology, Galway, Ireland

I. O' Connor
Galway-Mayo Institute of Technology, Galway, Ireland

J.P. Obbard
Qatar University, Doha, Qatar

S. Oberbeckmann
Leibniz Institute for Baltic Sea Research, Rostock, Germany; Leibniz Institute for Baltic Sea Research, Rostock, Germany

R. Obispo
Centro de Estudios de Puertos y Costas, Madrid, Spain; Division para la Proteccion del Mar., Madrid, Spain

R. Officer
Galway-Mayo Institute of Technology, Galway, Ireland

M. Ogonowski
Stockholm University, Stockholm, Sweden; Aquabiota Water Research AB (Aquabiota), Stockholm, Sweden

A. Orbea
Universidad del País Vasco/Euskal Herriko Unibertsitatea, Leioa, Spain

M. Ortlieb
Shimadzu Europa GmbH, Duisburg, Germany

A.M. Osborn
RMIT University, Bundoora, VIC Australia

P. Ostiategui-Francia
University of Las Palmas de Gran Canaria (ULPGC), Las Palmas de Gran Canaria, Spain

S. Outi
Finnish Environment Institute, Helsinki, Finland

T. Packard
Universidad de Las Palmas de Gran Canaria (ULPGC), Las Palmas de Gran Canaria, Spain

S. Pahl
Plymouth University, Plymouth, United Kingdom

A. Palatinus
Institute for Water of the Republic of Slovenia, Ljubljana, Slovenia

A. Palmqvist
Roskilde University, Roskilde, Denmark

P. Pannetier
University of Bordeaux, Talence, France

C. Panti
University of Siena, Siena, Italy

E. Parmentier
University of Liege, Liège, Belgium

P. Pasanen
University of Eastern Finland, Kuopio, Finland

T. Patarnello
Università di Padova, Padova, Italy

C. Pattiaratchi
The University of Western Australia (CEME), Crawley, WA, Australia

M. Pauletto
Università di Padova, Padova, Italy

M. Paulus
Trier University, Trier, Germany

K. Pavlekovsky
Sea Education Association, Woods Hole, MA, United States

H.B. Pedersen
Plastic Change, Hellerup, Denmark

M.-L. Pedrotti
CNRS-LOV, Villefranche sur Mer, France; Université Pierre et Marie Curie (UPMC), Paris, France

I. Peeken
Alfred Wegener Institute, Helmholtz Centre for Polar and Marine Research (AWI), Bremerhaven, Germany

D. Peeters
Institut Français de Recherche pour l'Exploitation de la Mer (IFREMER), Plouzané, France

E. Peeters
Wageningen University (WUR), Wageningen, Netherlands

D. Pellegrini
ISPRA, Italian National Institute for Environmental Protection and Research, Livorno, Italy

J.A. Perales
University of Cadiz, Cadiz, Spain

E. Perez
Université Paul Sabatier, Toulouse, France

V. Perz
Austrian Centre of Industrial Biotechnology, ACIB, Tulln, Austria

S. Petit
Université Pierre et Marie Curie (UPMC), Paris, France

M. Pflieger
University of Nova Gorica, Nova Gorica, Slovenia

C.K. Pham
Universidade dos Açores, Horta, Portugal; MARE – Marine and Environmental Sciences Centre, Horta, Portugal; IMAR – Instituto do Mar, Horta, Portugal

K. Philippe
Université de Perpignan, Perpignan, France

V. Piazza
Institute of Marine Science (ISMAR), Genova, Italy

M. Pinto
University of Vienna, Vienna, Austria

O. Planells
Marine Sciences For Society

M. Plaza
Centro de Estudios de Puertos y Costas, Madrid, Spain; Division para la Proteccion del Mar., Madrid, Spain

O. Pompini
Central Environmental Laboratory, EPFL, Lausanne, Switzerland

A. Potthoff
Fraunhofer Gesellschaft, IKTS, Dresden, Germany

L. Prades
University of the West of Scotland, Paisley, Scotland

S. Primpke
Alfred Wegener Institute, Helmholtz Centre for Polar and Marine Research (AWI), Bremerhaven, Germany; Alfred Wegener Institute Helmholtz Centre for Polar and Marine Research (AWI), Helgoland, Germany

M. Proietti
Universidade Federal do Rio Grande, Rio Grande, Brazil

G. Proskurowski
University of Washington, Seattle, WA, United States; MarqMetrix, Seattle, WA, United States

C. Puig
Consorci El Far, Barcelona, Spain

M. Pujo-Pay
Université Pierre et Marie Curie (UPMC), Paris, France

K. Pullerits
Tallinn University of Technology, Tallinn, Estonia

A.M. Queirós
Plymouth Marine Laboratory, Plymouth, United Kingdom

B. Quinn
University of the West of Scotland, Paisley, Scotland

E. Raimonds
University of Latvia, Riga, Latvia

J. Ramis-Pujol
Universitat Ramon Llull, Barcelona, Spain; ESADE Barcelona, Barcelona, Spain; ESADE Barcelona – Sant Cugat, Barcelona, Spain

R. Rascher-Friesenhausen
Hochschule Bremerhaven, Bremerhaven, Germany; Fraunhofer MEVIS, Bremen, Germany

E. Reardon
University of Exeter, Exeter, United Kingdom

F. Regoli
Università Politecnica delle Marche, Ancona, Italy

A.M. Reichardt
Leibniz Institute for Baltic Sea Research, Rostock, Germany

G. Reifferscheid
German Institute for Hydrology, Koblenz, Germany

K. Reilly
Plymouth Marine Laboratory, Plymouth, United Kingdom

J. Reisser
The Ocean Cleanup (TOC), ES Delft, Netherlands

I. Riba
Universidad de Cádiz, Cádiz, Spain

D. Ribitsch
Austrian Centre of Industrial Biotechnology, ACIB, Tulln, Austria; BOKU University of Natural Resources and Life Sciences, Vienna, Tulln, Austria

E. Rinnert
Institut Français de Recherche pour l'Exploitation de la Mer (IFREMER), Plouzané, France

N. Rios
OMA – Observatório do Mar dos Açores, Horta, Portugal

S.E. Rist
Technical University of Denmark, Kongens Lyngby, Denmark

M.M. Rivadeneira
Centro de Estudios Avanzados en Zonas Áridas (CEAZA), Coquimbo, Chile

G. Rivière
Agence nationale de sécurité sanitaire de l'alimentation, de l'environnement et du travail (Anses), Maisons-Alfort, France

J. Robbens
Institute for Agricultural and Fisheries Research, Merelbeke, Belgium

C.J.R. Robertson
Wellington, New Zealand

V. Rocher
SIAAP (syndicat interdépartemental pour l'assainissement de l'agglomération parisienne), Colombes, France

C.M. Rochman
University of California, Davis, CA, United States

M. Rodrigues
National Laboratory for Civil Engineering, Lisbon, Portugal

Y. Rodriguez
OMA – Observatório do Mar dos Açores, Horta, Portugal

A. Rodríguez
Universitat de Barcelona, Barcelona, Spain

G. Rodríguez
University of Las Palmas de Gran Canaria (ULPGC), Las Palmas de G.C., Spain; Institute of Environmental and Natural Resources Study (i-UNAT), ULPGC, Las Palmas de G.C., Spain

J.R.B. Rodríguez
Instituto Tecnológico de Canarias, Las Palmas, Spain

S. Rodríguez
Oficina de la Reserva de la Biosfera de Lanzarote, Lanzarote (Canary Islands), Spain

Y. Rodríguez
OMA – Observatório do Mar dos Açores, Horta, Portugal

E. Rogan
University College Cork, Cork, Ireland

E. Rojo-Nieto
University of Cadiz, Cadiz, Spain

T. Romeo
ISPRA, Italian National Institute for Environmental Protection and Research, Milazzo, Italy

P.S. Ross
Vancouver Aquarium Marine Science Centre, Vancouver, BC, Canada

A. Roveta
Istituto di Scienze Marine, Consiglio Nazionale delle Ricerche (ISMAR-CNR), Genova, Italy

S.J. Rowland
Plymouth University, Plymouth, United Kingdom

N.A. Ruckstuhl
c-o-u-p.org

A-C. Ruiz-Fernández
Instituto de Ciencias del Mar y Limnología, UNAM, Mazatlan, Mexico

L.F. Ruiz-Orejón
Centre d'Estudis Avançats de Blanes (CEAB-CSIC), Blanes-Girona, Spain

J. Runge
Gulf of Maine Research Institute, Portland, ME, United States

M. Russell
Marine Scotland – Science, Scottish Government, Aberdeen, Scotland

C. Saavedra
Instituto Español de Oceanografía, Vigo, Spain

R. Saborowski
Alfred Wegener Institute Helmholtz Centre for Polar and Marine Research (AWI), Bremerhaven, Germany

B.E. Sahin
Recep Tayyip Erdogan University (RTEU), Rize, Turkey

S. Sailley
Plymouth Marine Laboratory, Plymouth, United Kingdom

K. Sakaguchi-Söder
Technical University of Darmstadt, Darmstadt, Germany

I. Salaverria
Norwegian University of Science and Technology, Trondheim, Norway

A. Sánchez-Arcilla
Universitat Politècnica de Catalunya, Barcelona, Spain

J. Sánchez-Nieva
University of Cadiz, Cadiz, Spain

W. Sanderson
Heriot-Watt University (HWU), Edinburgh, Scotland

J.J. Santana-Rodríguez
Universidad de Las Palmas de Gran Canaria, Las Palmas de Gran Canaria, Spain

S. Santana-Viera
Universidad de Las Palmas de Gran Canaria, Las Palmas de Gran Canaria, Spain

M.B. Santos
Instituto Español de Oceanografía, Vigo, Spain

M.R. Santos
DRAM – Direção Regional dos Assuntos do Mar, Horta, Portugal

M.R. Sanz
Instituto Tecnológico de Canarias, Las Palmas, Spain

R. Sardá
Centre d'Estudis Avançats de Blanes (CEAB-CSIC), Blanes-Girona, Spain

H. Savelli
Food and Agricultural Organization of the United Nations (FAO), Rome, Italy

R. Schoeneich-Argent
James Cook University (JCU), Townsville, QLD, Australia; Australian Institute of Marine
Science, Townsville, QLD, Australia; Carl von Ossietzky University Oldenburg, Wilhelmshaven,
Germany

B.M. Scholz-Böttcher
University of Oldenburg, Oldenburg, Germany

F. Sciacca
Race For Water Foundation (R4W) Lausanne, Switzerland

R.P. Scofield
Canterbury Museum, Christchurch, New Zealand

O. Seatala
Finnish Environment Institute, Helsinki, Finland

M. Selenius
University of Eastern Finland (UEF), Kuopio, Finland

R. Sempere
Aix-Marseille Université, Marseille, France

Y. Senturk
Recep Tayyip Erdogan University (RTEU), Rize, Turkey

H. Serge
Université de Perpignan, Perpignan, France

Y. Shashoua
The National Museum of Denmark, København K, Denmark

P. Sherman
Imperial College London, London, United Kingdom

C. Sick
Plastic Change, Hellerup, Denmark

D. Siegel
University of California, Santa Barbara, CA, United States

J.P. Sierra
Universitat Politècnica de Catalunya, Barcelona, Spain

F. Silva
New University of Lisbon, Lisbon, Portugal

C. Silvestri
ISPRA, Italian National Institute for Environmental Protection and Research, Roma, Italy

G. Sintija
University of Latvia, Riga, Latvia

O. Sire
Université Bretagne Sud, Lorient, France

B. Slat
The Ocean Cleanup (TOC), ES Delft, Netherlands

A. Smit
University of the Western Cape, Bellville, South Africa

P. Sobral
Universidade Nova de Lisboa, Monte de Caparica, Portugal; NOVA.ID FCT, Caparica, Portugal

J. Sorvari
University of Eastern Finland, Kuopio, Finland

Z. Sosa-Ferrera
Universidad de Las Palmas de Gran Canaria, Las Palmas de Gran Canaria, Spain

M.G. Sotillo
Ente Público Puertos del Estado, Madrid, Spain

P. Soudant
Institut Universitaire Européen de la Mer (IUEM), Plouzané, France; UBO-CNRS-LPO, Brest, France

L. Speidel
Alfred Wegener Institute Helmholtz Centre for Polar and Marine Research (AWI), Helgoland, Germany

D.J. Spurgeon
Centre for Ecology and Hydrology, Wallingford, United Kingdom

M.K. Steer
Plymouth Marine Laboratory, Plymouth, United Kingdom; University of Plymouth, Plymouth, United Kingdom

C.C. Steindal
University of Oslo, Oslo, Norway

R. Stifanese
Istituto di Scienze Marine, Consiglio Nazionale delle Ricerche (ISMAR-CNR), Genova, Italy

A. Štindlová
Universidad de Las Palmas de Gran Canaria, Las Palmas de Gran Canaria, Spain

B. Stjepan
Finnish Environment Institute, Helsinki, Finland

L. Stuurman
Wageningen University (WUR), Wageningen, Netherlands

G. Suaria
CNR-ISMAR, La Spezia, Italy

C.G. Suazo
Justus Liebig University Giessen, Giessen, Germany

A. Sureda
University of Balearic Islands, Palma de Mallorca, Spain

C. Surette
Marine Sciences For Society; Université de Moncton, Moncton, NB, Canada

C. Svendsen
Centre for Ecology and Hydrology, Wallingford, United Kingdom

K. Syberg
Roskilde University, Roskilde, Denmark

Z. Tairova
Aarhus Universitet, Aarhus C, Denmark

J. Talvitie
Aalto University, Aalto, Finland

B. Tassin
University of Paris-Est, Créteil, France

M. Tazerout
Ecole des Mines de Nantes, Nantes, France

M.B. Tekman
Alfred Wegener Institute, Helmholtz Centre for Polar and Marine Research (AWI), Bremerhaven, Germany; Alfred Wegener Institute, Helmholtz Centre for Polar and Marine Research (AWI), Helgoland, Germany

A. ter Halle
Université Paul Sabatier, Toulouse, France

M. Thiel
Universidad Católica del Norte, Coquimbo, Chile; Millennium Nucleus of Ecology and Sustainable Management of Oceanic Island (ESMOI), Coquimbo, Chile; Centro de Estudios Avanzados en Zonas Áridas (CEAZA), Coquimbo, Chile

K.V. Thomas
Norwegian Institute for Water Research (NIVA), Oslo, Norway

R.C. Thompson
Plymouth University, Plymouth, United Kingdom

T. Tinkara
National Institute of Biology, Piran, Slovenia

V. Tirelli
OGS - National Institute of Oceanography and Experimental Geophysics, Trieste, Italy

P. Tomassetti
ISPRA, Italian National Institute for Environmental Protection and Research, Roma, Italy

E. Toorman
KU Leuven, Leuven, Belgium

J. Toppe
Food and Agricultural Organization of the United Nations (FAO), Rome, Italy

A. Tornambè
ISPRA, Italian National Institute for Environmental Protection and Research, Roma, Italy

R. Torres
Plymouth Marine Laboratory, Plymouth, United Kingdom

M.E. Torres-Padrón
Universidad de Las Palmas de Gran Canaria, Las Palmas de Gran Canaria, Spain

A.J. Underwood
University of Sydney, Sydney, NSW, Australia

M. Urbina
University of Exeter, Exeter, United Kingdom; Universidad de Concepción, Concepción, Chile

A. Usategui-Martín
University of Las Palmas de Gran Canaria (ULPGC), Las Palmas de Gran Canaria, Spain

R. Usta
Recep Tayyip Erdogan University (RTEU), Rize, Turkey

L. Valdés
IOC-UNESCO, Paris, France

A. Valente
University of Lisbon, Lisbon, Portugal

T. Valentina
National Institute of Biology, Piran, Slovenia

K. van Arkel
Race For Water Foundation (R4W) Lausanne, Switzerland

C. Van Colen
Ghent University, Gent, Belgium

N. Van Der Hal
University of Haifa, Haifa, Israel

J.A. van Franeker
Institute for Marine Research and Ecosystem Studies (IMARES), Texel, The Netherlands; Institute for Marine Resources and Ecosystem Studies (IMARES), Den Helder, Netherlands

L. Van Herwerden
James Cook University (JCU), Townsville, QLD, Australia

M. Van Loosdrecht
Delft University of Technology (TU Delft), BC Delft, Netherlands

A. van Oyen
CARAT GmbH, Bocholt, Germany

F. Vandeperre
MARE – Marine and Environmental Sciences Centre, Horta, Portugal; IMAR – Instituto do Mar, Horta, Portugal; Université de Versailles SQY, Guyancourt, France; Marine Sciences For Society

J-P. Vanderlinden
Université de Versailles SQY, OVSQ, CEARC, Guyancourt, France

D. Vani
ISPRA, Italian National Institute for Environmental Protection and Research, Roma, Italy

L. Vasconcelos
New University of Lisbon, Lisbon, Portugal

D. Vega-Moreno
University of Las Palmas de Gran Canaria (ULPGC), Las Palmas de Gran Canaria, Spain

A. Ventero
Instituto Español de Oceanografía, Madrid, Spain

A.D. Vethaak
VU Amsterdam, Amsterdam, The Netherlands; Deltares, Marine and Coastal Systems, Delft, The Netherlands

A. Vianello
Institute for the Dynamics of Environmental Processes and National Research Council (IDPA-CNR), Padova, Italy

M. Vicioso
Marine Sciences For Society

L.R. Vieira
University of Porto, Porto, Portugal; CIIMAR/CIIMAR-LA, Interdisciplinary Centre of Marine and Environmental Research, Porto, Portugal

M.K. Viršek
Institute for Water of the Republic of Slovenia, Ljubljana, Slovenia

M. Vos
University of Exeter, Exeter, United Kingdom

M. Wahl
GEOMAR Helmholtz Centre for Ocean Research Kiel, Kiel, Germany

N. Wallace
NOAA Office of Response and Restoration, Silver Spring, MD, United States

A. Walton
University of Exeter, Exeter, United Kingdom; Centre for Ecology and Hydrology, Wallingford, United Kingdom

J.J. Waniek
Leibniz Institute for Baltic Sea Research, Rostock, Germany

A. Watts
University of Exeter, Exeter, United Kingdom

L. Webster
Marine Scotland – Science, Scottish Government, Aberdeen, Scotland

C. Wesch
Trier University, Trier, Germany

E. Whitfield
Moores University, Liverpool, United Kingdom

A. Wichels
Alfred Wegener Institute Helmholtz Zentrum für Polar und Meeresforschung (AWI), Helgoland, Germany

A.M. Wieczorek
National University of Ireland Galway, Galway, Ireland

C. Wilcox
CSIRO Oceans and Atmosphere Flagship, Hobart, TAS, Australia

R.J. Williams
Centre for Ecology and Hydrology, Wallingford, United Kingdom

P. Wong-Wah-Chung
Aix-Marseille Université, Aix-en-Provence, France

S. Wright
University of Exeter, Exeter, United Kingdom

K.J. Wyles
Plymouth Marine Laboratory, Plymouth, United Kingdom; Plymouth University, Plymouth, United Kingdom

R. Young
Plymouth University, Plymouth, United Kingdom

M. Yurtsever
Sakarya University, Sakarya, Turkey

U. Yurtsever
Sakarya University, Sakarya, Turkey

L. Zada
VU Amsterdam, Amsterdam, The Netherlands

N.P. Zamani
Bogor Agricultural University, Bogor, Indonesia

G. Zampetti
LEGAMBIENTE, Roma, Italy

Printed in the United States
By Bookmasters